Springer Series in Advanced Manufacturing

The **Springer Series in Advanced Manufacturing** includes advanced textbooks, research monographs, edited works and conference proceedings covering all major subjects in the field of advanced manufacturing.
The following is a non-exclusive list of subjects relevant to the series:

1. Manufacturing processes and operations (material processing; assembly; test and inspection; packaging and shipping).
2. Manufacturing product and process design (product design; product data management; product development; manufacturing system planning).
3. Enterprise management (product life cycle management; production planning and control; quality management).

Emphasis will be placed on novel material of topical interest (for example, books on nanomanufacturing) as well as new treatments of more traditional areas.

As advanced manufacturing usually involves extensive use of information and communication technology (ICT), books dealing with advanced ICT tools for advanced manufacturing are also of interest to the Series.

Springer and Professor Pham welcome book ideas from authors. Potential authors who wish to submit a book proposal should contact Anthony Doyle, Executive Editor, Springer, e-mail: anthony.doyle@springer.com.

More information about this series at http://www.springer.com/series/7113

Weidong Li · Yuchen Liang · Sheng Wang
Editors

Data Driven Smart Manufacturing Technologies and Applications

Springer

Editors
Weidong Li
Faculty of Engineering, Environment
and Computing
Coventry University
Coventry, UK

Yuchen Liang
Faculty of Engineering, Environment
and Computing
Coventry University
Coventry, UK

Sheng Wang
Faculty of Engineering, Environment
and Computing
Coventry University
Coventry, UK

ISSN 1860-5168 ISSN 2196-1735 (electronic)
Springer Series in Advanced Manufacturing
ISBN 978-3-030-66848-8 ISBN 978-3-030-66849-5 (eBook)
https://doi.org/10.1007/978-3-030-66849-5

This Springer imprint is published by the registered company Springer Nature Switzerland AG
The registered company address is: Gewerbestrasse 11, 6330 Cham, Switzerland

Preface

The advance in the information and communication technologies, such as the Internet of Things, industrial sensors, sensor networks, etc., have impacted manufacturing profoundly. With the technologies, data generated from modern manufacturing systems is experiencing explosive growth, which has reached over 100 EB annually. Manufacturing data contains rich knowledge and offers a tremendous opportunity in the transformation of the conventional manufacturing paradigm to the smart manufacturing paradigm.

In the past, physics-based methodologies have been actively investigated for resolving manufacturing problems. However, it is challenging to developing accurate physical mechanisms for analyzing and predicting complex physical phenomena in manufacturing systems preciously in an efficient and cost-effective means. With the proliferation of data acquired from manufacturing and the significant increase of computational power, there is a need of developing artificial intelligence and advanced big data analytics to complement physical science to handle manufacturing data in supporting smart decision-making. For instance, in recent years, the developments of innovative machine learning algorithms especially deep learning algorithms are dramatic. The algorithms have been widely applied to mine knowledge from manufacturing data to facilitate various smart manufacturing applications. That is, smart manufacturing aims to take advantage of the state-of-the-art artificial intelligent technologies to enhance the adaptability, flexibility, robustness, and sustainability of physical manufacturing processes. Meanwhile, with the technologies, effectiveness and intelligence in manufacturing decision-making will be significantly improved. These methodologies and algorithms can be summarized as data driven manufacturing technologies, which will reinforce the competitiveness of manufacturing companies to address increasingly dynamic markets and sophisticated customer needs.

Owing to the importance and rapid progress of data driven smart manufacturing, it is essential to make a timely update on this subject. With this aim, this book reports the relevant applied research conducted by the authors in recent years in some important manufacturing applications, including intelligent diagnostics and prognostics on manufacturing processes, sustainable manufacturing, prediction of cutting tool life, heavy-duty machining applications, prediction of strength of adhesive bonded joints, human–robot cooperation, etc. Majority of the first authors in chapters are Ph.D. students, whose works have been supported by some important research projects sponsored by the European Commission, the Innovate UK (the UK industrial funding agency), the Institute of Digital Engineering of the UK, and manufacturing companies from the UK and China. Meanwhile, some works have been supported by the National Natural Science Foundation of China (project no. 51975444) and international cooperative projects between the UK and China. In order to make this book to be a systematic reference, some brief theoretical foundations of deep learning algorithms, as well as the state-of-the-art intelligent algorithms and practical cases on the aforementioned manufacturing applications, are covered in the book. Case studies in this book illustrate design details of intelligent algorithms and methodologies for various manufacturing applications, in a bid to provide a useful reference to readers.

This book supplements the Springer books "Cloud Manufacturing" and "Research on Intelligent Manufacturing" by providing technical discussions and implementation details of data driven manufacturing technologies and applications. We believe that the book offers a valuable resource for researchers in the smart manufacturing communities, as well as practitioners, engineers, decision-makers in industry and all those interested in smart manufacturing and Industry 4.0.

Dr. Weidong Li
Professor
Wuhan University of Technology
Wuhan, China

Professor
Coventry University
Coventry, UK

Dr. Yuchen Liang
Lecturer
Coventry University
Coventry, UK

Dr. Sheng Wang
Researcher
Coventry University
Coventry, UK

Contents

About the Editors

Prof. Weidong Li is a full professor in manufacturing, Coventry University (UK). Professor Li has more than 20 years of experience in computer-aided design, manufacturing informatics, smart manufacturing, and sustainable manufacturing. His research has been sponsored by a number of research and development projects from the UK EPSRC, European Commission, and European industries such as Jaguar Land Rover, Airbus, Rolls-Royce, Sandvik, etc. In the research areas, he has published four books and around 200 research papers in international journals and conferences.

Dr. Yuchen Liang is a lecturer from Coventry University. Dr. Liang has gotten his Ph.D. degree from Coventry University from Automotive and Mechanical Engineering. His research areas are data driven smart manufacturing and cyber-physical manufacturing systems. His research works have been sponsored by the European Commission and the Innovate UK.

Dr. Sheng Wang is a senior researcher in manufacturing, Coventry University, UK. Dr. Wang got her Ph.D. degree from Queen Mary University of London from Computer Science and Electronic Engineering. In the past five years, Dr. Wang has participated in a number of European Commission-sponsored projects in smart manufacturing.

About the Editors

Introduction

W. D. Li, Y. C. Liang, and S. Wang

1 Smart Manufacturing Trend

In the past century, the manufacturing industry has undergone significant paradigm shifts, from mass production (Ford assembly line) (1900s) to lean manufacturing (Toyota production system) (1960s), flexible manufacturing (mass customization) (1980s), reconfigurable manufacturing (1990s), collaborative manufacturing (2000s), and smart manufacturing (2010s) [1–3]. The development trend of the global manufacturing industry is shown in Fig. 1.

In 2010, Germany proposed the framework of Industry 4.0 in transforming factories towards smart manufacturing [4]. Industry 4.0 interprets major manufacturing paradigm shifts from the perspective of prominent breakthrough of enabling technologies. The 1st industrial revolution was signified by the invention of the steam engine, which empowered mechanical machines to expand manufacturing capabilities. The emergence of electrical energy indicated the 2nd industrial revolution, through which mass production lines were electrified contributing to significantly improved productivity, product quality repeatability and process control. Computerized numerical controlling (CNC) machines and programming logic controllers were some important technologies deployed in factories. The 3rd industrial revolution witnessed the wide adoption of electronics, computers, programming languages, and various industrial software systems (computer-aided design (CAD), computer-aided manufacturing (CAM), computer-aided engineering (CAE), manufacturing execution systems (MES), etc.). Design and manufacturing processes were digitalized, and close-loop control and programable/re-programable automation in manufacturing

W. D. Li (✉)
School of Logistics Engineering, Wuhan University of Technology, Wuhan, China
e-mail: weidong.li@coventry.ac.uk

W. D. Li · Y. C. Liang · S. Wang
Faculty of Engineering, Environment and Computing, Coventry University, Coventry, UK

W. Li et al. (eds.), *Data Driven Smart Manufacturing Technologies and Applications*, Springer Series in Advanced Manufacturing,
https://doi.org/10.1007/978-3-030-66849-5_1

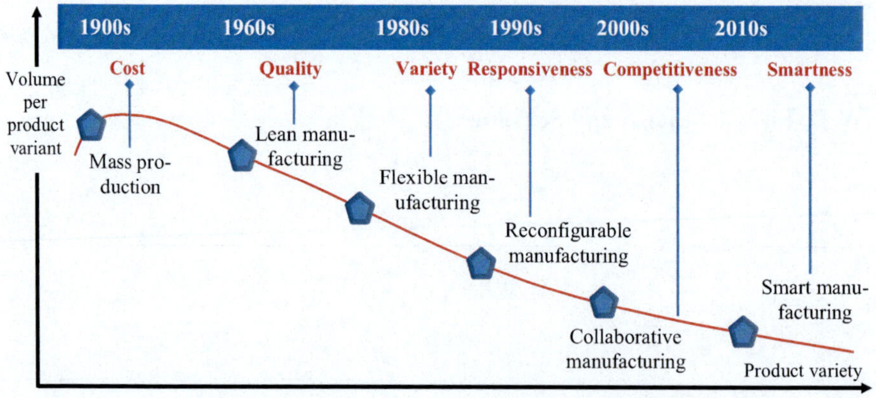

Fig. 1 Significant paradigm shifts of the global manufacturing industry

were realized. Lean, flexible, reconfigurable, customable and collaborative production lines were subsequently implemented in factories. The 4th industrial revolution was characterized by the proliferate use of industrial sensors, the Internet of Things, machine learning, deep learning, big data analytics and cloud computing to establish smart manufacturing. During the era, design of cyber-physical systems for smart factories and processing of the vast amount of manufacturing data for smart decision making were central topics. The review of Industry 4.0 and relevant technological development can be found in [5,6] (illustrated in Fig. 2).

Apart from Industry 4.0, in recent years, strategic roadmaps of smart manufacturing have been recently developed by some major industrialized countries to take advantage of the state-of-the-art information technologies. In 2011, the Smart Manufacturing Leadership Coalition (SMLC) in the U.S. created a systematic framework of smart manufacturing to be implemented in American factories [6]. In 2014, the

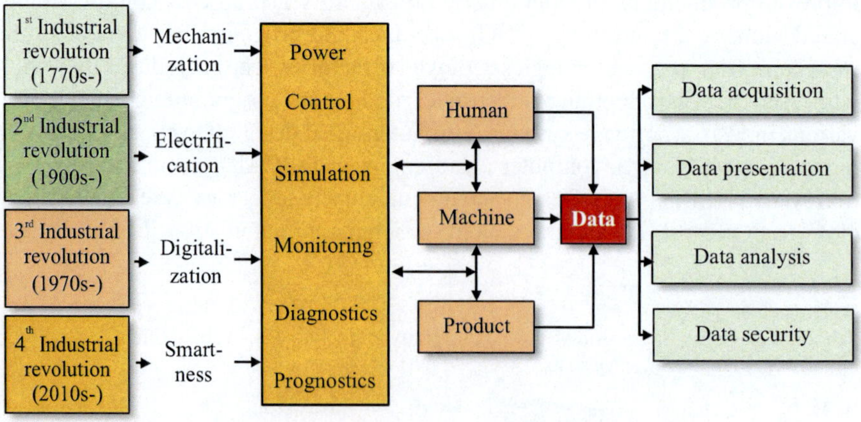

Fig. 2 Industry 4.0 and key technologies supporting manufacturing

Manufacturing Industry Innovation 3.0 strategy was launched in Korea to realize the Korean strategy of smart manufacturing. The plan "China Manufacturing 2025", initialized in China in 2015, aims to promote advanced manufacturing in China [7]. The National Institute of Standards and Technology (NIST) defined smart manufacturing as "fully-integrated, collaborative manufacturing systems that respond in real time to meet changing demands and conditions in the factory, in the supply network and in customer needs" [6]. According to the factories of the future roadmap of European Union, European factories should focus on key priorities under a common headline "co-creation through manufacturing eco-systems" [8], including: (1) responsible and smart manufacturing systems; (2) manufacturing with zero-environmental impact; (3) customer-driven value networks; (4) parallel product and manufacturing engineering; (5) human-driven innovation.

2 Data Driven Smart Manufacturing

Physics-based methodologies have been actively investigated for manufacturing problems. For example, physical mechanisms of manufacturing elements are analyzed based on the Finite Element Method (FEM), Finite Difference Method (FDM), or Finite Difference Element Method (FDEM). However, it is challenging to developing accurate physical mechanisms for analyzing and predicting complex physical phenomena in manufacturing systems preciously. With the proliferation of data acquired from manufacturing and the significant increase of computational power, there is a need of developing artificial intelligence and advanced data analytics to complement physical science to handle manufacturing data for smart decision making. In recent Gartner's reports for strategic technology trends in the world, artificial intelligence has been listed as a top strategic technology or the foremost enabling technology in supporting smart manufacturing application [9].

The artificial neural network (ANN) is one of the most important progresses in the artificial intelligence development. The initial design of ANN can be traced back in 1940s, when simple ANN models were proposed to simulate how neurons work in the human brain for linear optimization [10]. The models were implemented using electrical circuits. In 1970s, the back propagation (BP) algorithm was designed for ANN to solve non-linear problems in complex neural networks [11]. As more sensors have been installed in factories' shop floors, manufacturing has been getting smarter, and data captured throughout manufacturing life cycles have been growing exponentially. Accumulated data in manufacturing are often considered as big data since they can be described in four Vs: Volume (there are a large amount of collected data), Variety (collected data are in different forms or types), Velocity (data change or are updated frequently), and Veracity (it refers to the biases, noise and abnormality in collected data). ANN, however, could not converge in resolving such big data. To tackle this issue, in the 1990s, deep learning models, which are extended versions of ANN, were proposed.

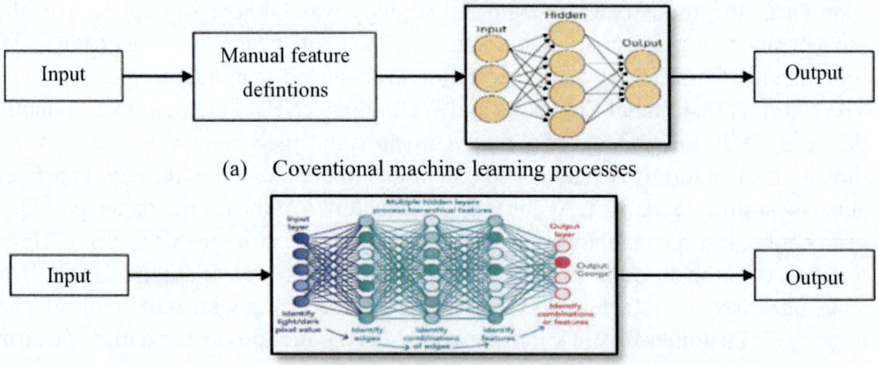

(a) Coventional machine learning processes

(b) Deep learning processes

Fig. 3 Comparison between conventional machine learning and deep learning [12]

Deep learning models were considered as breakthrough techniques and demonstrated outstanding performance in large-volume data processing in the areas of speech recognition, image recondition, natural language processing, multimodal image-text, games (e.g., Alphago), and various engineering and manufacturing applications. In comparison with ANN, support vector machine (SVM) and Boltzmann machine, deep learning models allow processing highly non-linear and complex data towards feature abstraction via a cascade of multiple layers and leaning mechanisms. It overcomes the laborious process of a conventional machine learning model to handcraft optimum feature representations and extract features from data. The different processes between deep learning models and the conventional machine learning models is illustrated in Fig. 3.

Deep learning offers a great potential to boost data driven smart manufacturing, from product quality inspection to manufacturing process health management, and manufacturing optimization in terms of improved energy saving, productivity and quality performance, etc. For example, manufacturing systems are subject to various failures or defects caused by excessive load, fracture, overheating, corrosion and wear. Diagnostics and prognostics are core functions of manufacturing process health management that allow manufacturers to optimize the conditions and performance of manufacturing systems. A schematic diagram for some potential smart manufacturing applications based on deep learning models is shown in Fig. 4.

3 Typical Deep Learning Models

3.1 Some Development of Deep Learning Models

Deep learning is a branch of machine learning that is an extended version of ANN in mimicking the human brain. A deep learning model processes data through multiple

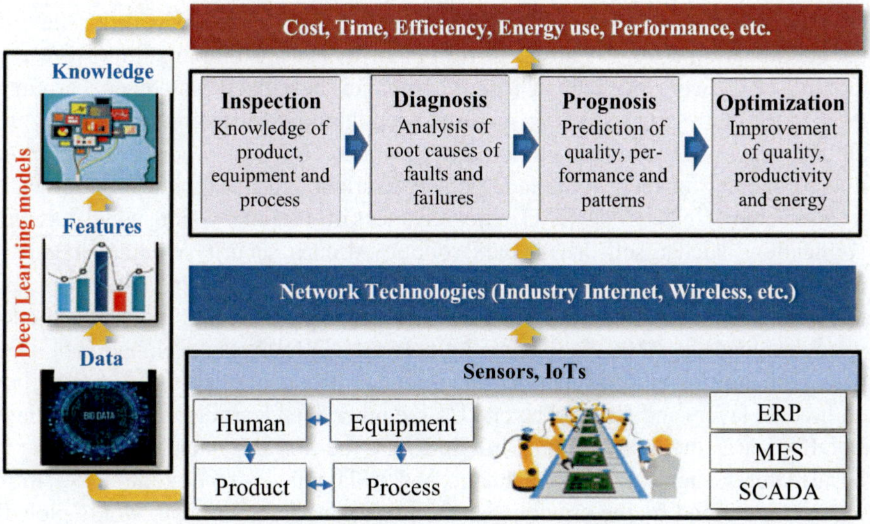

Fig. 4 Smart manufacturing and deep learning models

(deeper) layers of neural network, each of which passes features recognized from data into the next layer. Based on deep learning models, features from a vast amount of data can be extracted and represented autonomously rather than the process of a conventional machine learning model by which explicit engineered features are used to perform for this process. That is, deep learning is an end-to-end learning model to integrate feature learning and model construction with the minimum human inference. The architecture of a deep learning model is composed of multiple hidden layers to conduct multi-level and non-linear computational operations. In each layer, features are extracted from the original input and transferred into more abstracted features in higher layers to identify complicated inherent relationships in the original data. These abstracted features are then input into a final layer to perform classification, clustering or regression tasks.

From the mid of 1990s, there are important progresses in deep learning development, and some typical deep learning models were proposed. In 1995, the recurrent neural network (RNN) was designed [13]. In RNN, the outputs from neurons are used as feedback to the neurons of the previous layer recurrently. The architecture of RNN is naturally suited to processing time-series data and other sequential data for prediction or control tasks.

In 1997, an improved version of RNN, namely the long short-term memory (LSTM), was developed to tackle the vanishing and exploding gradient problems in RNN to process complex time-series data [14].

In 1998, the convolutional neural network (CNN) was devised to facilitate images or other two-dimensional information, in which feature learning was achieved by stacking convolutional layers and pooling layers [15]. The success of a deep convolutional architecture called AlexNet in the 2012 ImageNet competition (the annual

Olympics of computer vision) was the shot heard round the world. Using the AlexNet, the classification error was dropped from 26 to 15%, an astounding improvement at the time [16]. Ever since then, a host of companies have been using deep learning at the core of their services. Facebook uses deep learning for their automatic tagging algorithms, Google for their photo search, Amazon for their product recommendations, Pinterest for their home feed personalization, and Instagram for their search infrastructure. CNN has powered major advances in computer vision, which has wide applications, such as self-driving cars, robotics, drones, security, medical diagnoses, and treatments for the visually impaired.

From 2000s, deep learning has gained wider popularity, and more new models were designed. In 2006, the deep belief network (DBN) was developed [17]. DBN uses probabilities and unsupervised learning to produce outputs. It contains both undirected layers and directed layers. Unlike other deep learning models, each layer in DBN learns the entire input, while in CNN, the first layers only filter inputs for basic features, such as edges of an image, and the later layers recombine all the simple features identified by the previous layers. DBN, on the other hand, works globally and regulates each layer in order. Greedy learning algorithms are used to pre-train DBN. This is a problem-solving approach that involves making the optimal choice at each layer in the sequence, fine-tuning the model parameters and eventually finding a global optimum.

In 2009, the deep Boltzmann machine was designed to have a bidirectional structure to learn ambiguous input, and the model parameters were optimized using layer-wise pre-training efficiently [18].

In 2014, the generative adversarial network (GAN) was devised to learn from a set of training data and generate new data with the same characteristics as the training data [19]. GAN consists of two neural networks, i.e., the generator and the discriminator, which compete against each other. The generator is trained to produce fake data, and the discriminator is trained to distinguish the generator's fake data from real examples. If the generator produces fake data that the discriminator can easily recognize as implausible, such as an image that is clearly not a face, the generator is penalized. Over time, the generator learns to produce more plausible examples. GAN has been used for some applications, such as image editing, animation model generation, prognostics and heath management in manufacturing.

Deep learning models are still in development. For example, to address issues of CNN, improved models were developed, including regions with CNN (R-CNN) (2014) [20], fast R-CNN (2015) [21], capsule networks (CAPSNet) [22], etc.

3.2 *Typical Deep Learning Algorithms*

A deep learning model is generally classified as a supervised and an unsupervised learning model. A supervised learning model is trained to classify data or predict outcomes accurately based on pre-defined (supervised) labeled datasets. In

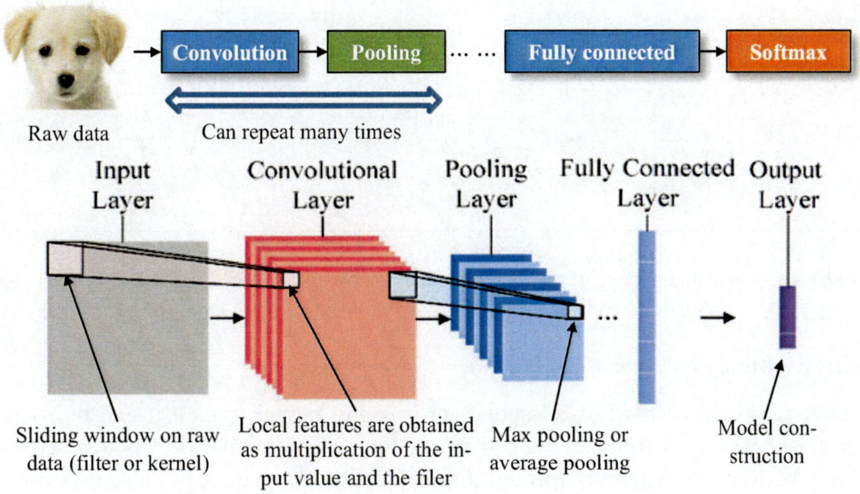

Fig. 5 A typical architecture of the convolutional neural network (CNN)

contrast, an unsupervised learning model analyzes and clusters unlabeled datasets autonomously by discovering similarities and differences in the datasets.

(1) Convolutional neural networks (CNN).

CNN is a supervised learning model that can be used to resolve classification, regression or clustering problems. A typical architecture of CNN is shown in Fig. 5. In CNN, features are learnt from an input raw dataset by stacking a series of convolutional and pooling layers. Learnt local features can be synchronized in a fully connected layer and then fed into a softmax layer for classification or regression. The convolutional layers convolve with raw input data using multiple kernel filters and generate invariant local features. The subsequent pooling layers extract the most significant features using pooling operations, such as max pooling or average pooling, to select the maximum or mean value of one region of the feature map as the most significant feature.

Furthermore, a fully connected layer is used to convert a two-dimensional feature map converted from the convolutional and pooling layers into a one-dimensional vector. The one-dimensional vector will be further fed into a softmax function for normalization.

In CNN training, there could be an "overfitting" issue, which refers to the phenomenon that CNN is over-sensitive to some small variations in the training dataset and cannot effectively reflect the features of the dataset. To address the issue, CNN design could also include a dropout layer to randomly deactivate some neurons in a given layer, thus forcing the network to adapt to learn with less-informative data.

Gradient based backpropagation is a popular algorithm to train CNN by minimizing the minimum mean squared error or a cross-entropy loss function against the labeled dataset.

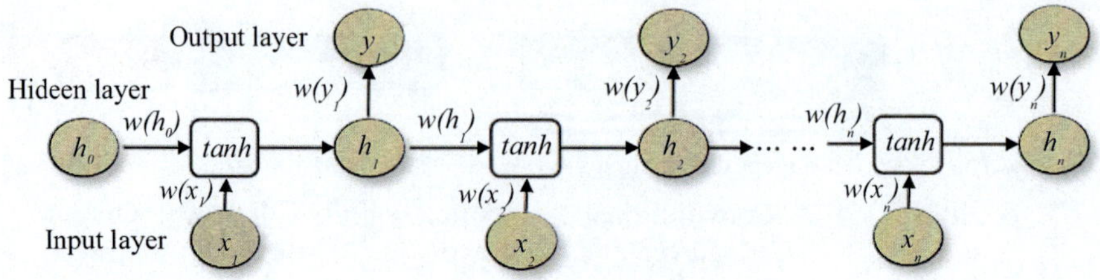

Fig. 6 A typical architecture of the recurrent neural network (RNN)

(2) Recurrent neural network (RNN).

RNN is primarily used as a supervised learning model though it can be used as an unsupervised learning model. It has unique characteristic of topology connections between the neurons formed directed cycles for sequence data. It is used in various applications, such as handwriting recognition, speech recognition, machine translation, character-level language modeling, image classification, and financial engineering. A typical architecture of RNN is shown in Fig. 6. RNN allows that information persists in hidden layers and captures previous states of a few time steps ago. Through such architecture design and learning mechanism, RNN has a memory that allows it to better recognize patterns in sequence data.

RNN is an extension of ANN by adding connections to feed the hidden layers of the previous state into themselves. That is, RNN works by iteratively updating a hidden state h_i $(i = 1,\ldots,n)$. At any given step *i*, firstly, the hidden state h_i is calculated using the previous hidden state h_{i-1} and the current input x_i; secondly, the output y_i is calculated using *hi* through a softmax function. Therefore, unlike ANN, RNN uses the past events to process the input vector at the current time rather than starting from scratch every time. This characteristic provides RNN an ability to memorize past events for future tasks, such as prediction and close-loop control.

The model training in RNN is performed by backpropagation through time (BPTT). BPTT is applied to calculate gradients for training.

(3) Long short-term memory (LSTM).

Due to the vanishing and exploding gradient problem using BPTT for model training, RNN cannot capture long-term dependencies so that it is difficult to handle long-term sequence data. To address the issue, LSTM was developed. Comparing with the single recurrent structure in RNN, each recurrent unit (cell) in LSTM is enhanced to include a forget gate, an input gate and an output gate. The design can enable each cell to adaptively capture long-term dependencies of different time scales. A typical architecture of RNN is shown in Fig. 7.

In LSTM, a forget gate is responsible for removing information from the cell state. The information that is no longer required for LSTM to understand things or the information that is of less importance is removed via multiplication of a filter. This can optimize the performance of LSTM.

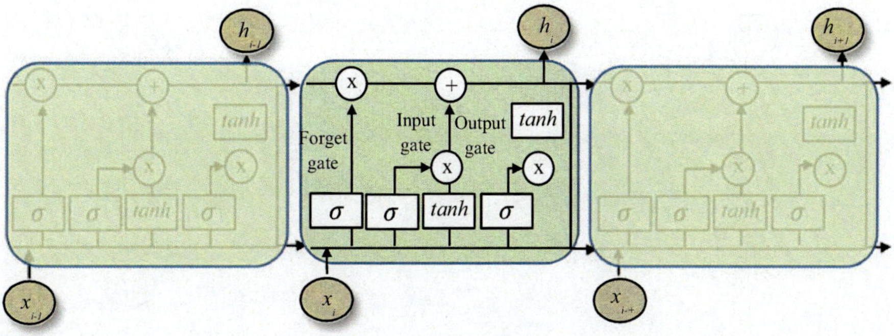

Fig. 7 A typical architecture of the long short-term memory (LSTM)

The input gate is responsible for the addition of information to the cell state via a three-step process, i.e., (i) regulating what values need to be added to the cell state by involving a sigmoid function, (ii) creating a vector containing all possible values that can be added to the cell state via the tanh function, (iii) multiplying the value of the regulatory value (the sigmoid function) to the created vector (the tanh function) and then adding this useful information to the cell state via an addition operation. Once this three-step process is completed, it ensures that only information added to the cell state is important and is not redundant.

In the output gate, useful information from the current cell state will be generated via a three-step process, i.e., (i) creating a vector after applying the tanh function to the cell state, thereby scaling the values to a preferred range, (ii) regulating the values that need to be output from the vector created above via a sigmoid function, (iii) multiplying the regulatory value to the vector created in (i), and sending it out as an output and also to the hidden state of the next cell.

(4) Auto encoder (AE).

Auto encoder (AE) is an unsupervised learning model exacting features from input data without label data. It mainly consists of two parts, i.e., encoder and decoder. A typical architecture of AE is illustrated in Fig. 8.

In AE, the encoder performs data compression especially in processing input with high dimensionality to map the input to a hidden layer. The decoder reconstructs the approximation of input. If the activation function in AE is a linear function and there are few hidden layers than the dimensionality of input, AE is similar to the principle component analysis (PCA). Otherwise, if the input is highly nonlinear, more hidden layers are required to construct AE.

Stochastic gradient descent (SGD) is often investigated to calculate the parameters of AE by minimizing the loss function of objectives in terms of the least square loss or cross-entropy loss.

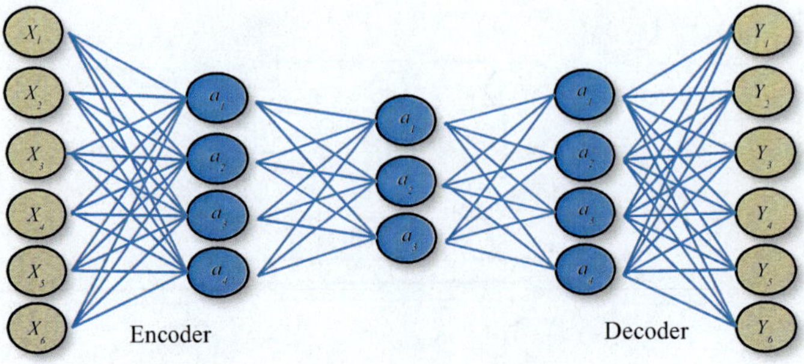

Fig. 8 A typical architecture of the auto encoder

3.3 *Further Discussions on Data Driven Smart Manufacturing*

Data is centric to data driven smart manufacturing, and the performance of deep learning is heavily dependent on the scale and quality of data. However, there are some issues in data collected from shop floors and it is challenging to implementing smart manufacturing effectively. Thus, optimization measurements and strategies are imperative.

- *High dimensional and multi-modality manufacturing data.* In manufacturing shop floors, multiple sensors (e.g., vibration, acoustic emission, electricity) are increasingly instrumented to reflect the different perspectives and stages of a manufacturing life cycle comprehensively. Deep learning models, however, are ineffective or even infeasible in processing the high dimensional, multi-modality and non-structured manufacturing data collected from the multiple sensors. Therefore, it is important to devise a data fusion strategy so that the size and the structure of the data will be optimized to improve the performance of applied deep learning models.
- *Imbalanced manufacturing data.* The class imbalance is a challenge in data driven smart manufacturing. That is, the class follows a highly skewed distribution in practical manufacturing, representing most data sample belong to few categories [11]. Thus, it is difficult to apply standard classification techniques to differentiating good parts from scraps. Appropriate measures such as class resampling, cost-sensitive training, and integration of boot strapping may be necessary for deep learning models to address class imbalance issues [22].
- *Redundant manufacturing data.* In manufacturing shop floors, a large number of sensors could be required to install in order to collect sufficient data for accurate modelling analysis. Nevertheless, sensors could be excessively deployed to generate redundant data if without an appropriate installation guidance provided. Data driven smart manufacturing could be severely hindered by the less efficient processes of collecting and handling large-volume data. To optimize the number of deployed sensors and minimize excessive data, it is worth integrating to hybridize

physics driven modelling to optimize the deployment of sensors to facilitate data driven smart manufacturing.

- *Dynamic manufacturing data.* Deep learning models are not built to learn incrementally and therefore susceptible to changing environments [11]. One of the reasons hindering deep learning models to be widely adopted by industries is the inadaptability of deep learning models to address the changing working conditions of customized manufacturing processes. For example, in practical manufacturing, it could be significantly time-consuming and expensive to re-collect a large amount of data and re-train a deep learning model previously trained for each new manufacturing condition. To enhance the adaptability of trained deep learning models on new conditions with small (limited) manufacturing data, deep transfer learning is adopted, where a maximum mean discrepancy (MMD) is defined and minimized between source (an original condition used to train the deep learning model) and target (a new condition) domains [23]. Thus, deep transfer learning could be a promising solution to enable deep learning models updating and adaptive throughout manufacturing life cycles dynamically.
- *Data analytics model selection.* Machine learning consists of a series of models. Each model has own characteristics and can tack different problems. In manufacturing, sometimes, it could be difficult to create a large amount of data to support deep learning models. For example, in a scenario of the intelligent trajectory generation of an industrial robot via learning from a human, demonstration data by the human are limited. In this case, deep learning models are not suitable. Other machine learning models designed for a small amount of data will be used.

In the rest of the book, industrial case studies and applications of data driven smart manufacturing, in the aspects of energy-efficient computerized numerical control (CNC) machining optimization, production health management, cutting tool life estimation, manufacturing thermal error compensation, human-robotic collaboration, etc., will be elaborated to illustrate the concepts, models and algorithms presented in this chapter.

References

1. Koren Y (2010) The global manufacturing revolution: product-process-business integration and reconfigurable systems. Wiley & Sons
2. Li WD, Ong SK, Nee AYC, McMahon CA (2007) Collaborative Product design and manufacturing methodologies and applications (Advanced Manufacturing Series). Springer
3. Li WD, Mehnen J (2013) Cloud manufacturing (Springer Series in Advanced Manufacturing). Springer
4. Industry 4.0: the fourth industrial revolution - guide to Industry 4.0. https://www.i-scoop.eu/industry-40-0/. Accessed 28 Sept 2020
5. Gao RX, Wang LH, Helu M, Teti R (2020) Big data analytics for smart factories of the future. CIRP Annal – Manufact Technol 69:668–692
6. Tao F, Qi Q, Liu A, Kusiak A (2018) Data-driven smart manufacturing. J Manufact Syst 48:157–169

7. SMLC: Smart Manufacturing Leadership Coalition. https://smartmanufacturingleadershipcoalition.org/. Accessed 29 Sept 2020
8. China Manufacturing 2025. https://www.cscc.it/upload/doc/china_manufacturing_2025_putting_industrial_policy_ahead_of_market_force[english-version].pdf. Accessed 29 Sept 2020
9. EFFRA: factories of the future roadmap. https://www.effra.eu/factories-future-roadmap. Accessed 29 Sept 2020
10. Gartner's top 10 strategic technology trend for 2020, https://www.gartner.com/smarterwithgartner/gartners-top-10-technology-trends-for-2020/. Accessed 27 Sept 2020
11. McCulloch WS, Pitts WH (1943) A logical calculus of the ideas immanent in nervous activity. Bull Math Biophys 5(4):115–133
12. Werbos PJ (1990) Backpropagation through time: what it des and how to do it. Proc IEEE 78(10):1550–1560
13. Wang J, Ma Y, Zhang L, Gao RX, Wu DZ (2018) Deep learning for smart manufacturing: Methods and applications. J Manufact Syst 48:144–156
14. Hihi SE, Bengio Y (1995) Hierarchical recurrent neural networks for long-term dependencies. Proceedings of the 8th International Conference on Neural Information Processing System. pp 493–499
15. Hochreiter S, Schmidhuber J (1997) Long short-term memory. Nerual Comput. 9(8):1735
16. Lecun Y, Bottou L, Bengio Y, Haffner P (1998) Gradient-based learning applied to document recognition. Proc IEEE 86(11):2278–2324
17. Krizhevsky A, Sutskever I, Hinton GE (2012) ImageNet classification with deep convolutional neural networks. Proceedings of the Advances in Neural Information Processing Systems. pp 1097–1105
18. Hinton GE, Salakhutdinov RR (2006) Reducing the dimensionality of data with neural network. Science 313(5786):504–507
19. Salakhutdinov RR, Hinton GE (2009) Deep Boltzmann machines. J Mach Learn Res 5(2):1967–2006
20. Goodfellow IJ, Pouget-Abadie J, Mirza M, Xu B, Warde-Farley D, Ozair S, Courville A, Bengio Y (2014) Generative adversarial nets. Proceedings of the 27th International Conference on Neural Information Processing Systems 2:2672–2680
21. Girshick R, Donahue J, Darrell T, Malik J (2014) Rich feature hierarchies for accurate object detection and semantic segmentation. https://arxiv.org/abs/1311.2524v5. Accessed 2 Oct 2020
22. Girshick R., 2015. Fast R-CNN. https://arxiv.org/pdf/1504.08083.pdf. Accessed 2 Oct 2020
23. Sabour S, Frosst N, Hinton GE (2017) Dynamic routing between capsules. https://arxiv.org/pdf/1710.09829.pdf. Accessed 3 Oct 2020

Fog Computing and Convolutional Neural Network Enabled Prognosis for Machining Process Optimization

Y. C. Liang, W. D. Li, X. Lu, and S. Wang

1 Introduction

During Computerized Numerical Control (CNC) machining processes, deviations on dimensional accuracy and surface roughness would lead to product defects or wastes. The root causes of deviations could be from poor tooling or abnormal equipment conditions, resulting in unsatisfactory precision, leaving scratch marks on machined components, generating severe vibration or high temperature during machining. Thus, it is significant to design prognosis systems to perform efficient analysis on condition data acquired from machines and processes in order to detect potential failures in early stages, leading to preventive maintenance and adaptive optimization [1]. Recently, cloud enabled prognosis systems have been developed by taking advantage of the computing and storage capabilities of cloud centers [2], as well as being reinforced by the state-of-the-art artificial intelligence (AI) algorithms to improve the accuracy and reliability of prognosis analysis [3]. Conversely, the performance of the systems is hindered by the high latency of data transfer between shop floors and the cloud. To overcome the limitation of the cloud model, a fog (From cOre to edGe) computing model based on the "divide and conquer" strategy has been researched [3–5]. Under the model, most of computing and intelligent reasoning can be conducted on edge devices locally for efficient analysis. Only necessary information is transferred to the cloud for intensive computation. Therefore, the bandwidth requirement is minimized, and the agility of detecting fault features during early stages is enhanced. Also, the data leakage issue pertaining to uploading all operation data into cloud centers is mitigated.

Y. C. Liang · W. D. Li (✉) · X. Lu · S. Wang
Faculty of Engineering, Environment and Computing, Coventry University, Coventry, UK
e-mail: weidong.li@coventry.ac.uk

W. D. Li
School of Logistics Engineering, Wuhan University of Technology, Wuhan, China

W. Li et al. (eds.), *Data Driven Smart Manufacturing Technologies and Applications*, Springer Series in Advanced Manufacturing,
https://doi.org/10.1007/978-3-030-66849-5_2

Though several fog enabled manufacturing systems have been recently reported [6–9], the systems are ineffective to meet the requirements of practical manufacturing processes. Research gaps and industrial requirements on fog enabled prognosis are summarized below:

In practical manufacturing environments, fault features are hidden in fluctuating signals and disturbed by non-Gaussian noises in the signals. The occurrence of faults is uncertain in long-term process execution, requiring effective AI algorithms (e.g., deep learning) to mine fault features from large-scale monitored data. The above processes generate intensive computation during prognosis analysis.

It is imperative to design a suitable fog enabled prognosis architecture to facilitate the intensive computation of related algorithms to support real-world applications. To leverage the capacities of fog devices effectively, it is vital to design pre-processing mechanisms to partition and de-noise monitored signals to facilitate the intensive computation of prognosis. Industrial trials/case studies will be essential to verify the advantage of the model to implement industrial artificial intelligence.

To address the above requirements, this chapter presents an innovative fog and deep learning enabled system for machining process prognosis and optimization. The functions and contributions of the system are summarized below:

Convolutional Neural Network (CNN) based prognosis is designed to achieve high detection rates for customized and varying machining processes. In this aspect, the novelty includes: (1) pre-processing mechanisms for the CNN to de-noise monitored data and enhance the computing efficiency effectively, and (2) suitable structures of the CNN designed to improve computation efficiency of prognosis.

To optimize data transfer, monitored data acquired on equipment are processed on a fog layer to identify faults efficiently using the trained CNN deployed locally. Intensive computations such as the training process of the CNN and re-scheduling optimization to address abnormal situations are carried out on a cloud layer.

The system was deployed into a UK machining company. Case studies were carried out to demonstrate the applicability of the system to manufacturing practices. With the deployed system, the energy and production efficiency were improved approximately for 29.25% and 16.50%. In comparison with cloud solutions, 70.26% reduction on bandwidth requirement, and 47.02% reduction of time spent for data transfer were achieved. To the best of our knowledge, it is the first time to develop a fog enabled prognosis system that is seamlessly integrated with deep learning algorithms and validated in practical machining processes.

2 Literature Survey

Generally, developed fog architectures are comprised of three layers: a terminal layer, a fog layer and a cloud layer [5]. The terminal layer is close to physical environment with the Internet of Thing (IoT) devices, such as sensors, smart readers, etc. The fog layer, which consists of Wi-Fi routers, gateways, etc., is usually located on the edge of the sensor network. A gateway has computing capabilities to process majority of acquired data with low latency, so that only the most useful data are sent to the cloud layer for complex computation. In recent years, there are increasingly reported research works of fog computing supported manufacturing. Wu et al. [7] proposed a fog framework for manufacturing optimization. The random forest algorithm was embedded to predict tool wear. Lin and Yang [8] developed a fog system to process data collected by sensors at logistics centers. The discrete monkey algorithm and genetic algorithm were applied to identify the best good-quality solutions efficiently. Wan et al. [9] designed a fog structure to optimize the energy consumption and workload of machines in a manufacturing shop floor. O'Donovan et al. [10] designed a fog architecture integrating machine learning algorithms. The execution time of both fog and cloud layers were analyzed to showcase the benefits of designing the fog layer. A particle swarm optimization algorithm was implemented on fog. Mohamed et al. [11] developed a fog structure to support multi-robot applications. However, the aforementioned systems in general lack validation based on complex industrial scenarios. Relevant analyses and results are mainly based on simulation.

In manufacturing, diagnosis and prognosis on manufacturing equipment and processes have been an active research field. In recent years, owing to the rapid progress of artificial intelligence, it has become an active trend to apply deep learning and Big Data analytics into this research field (several recent surveys were reported in [1, 2, 12–14]). Signals from manufacturing systems, such as current, vibration or acoustic emission, were continuously collected, and Big Data analyses were carried out in the time domain, the frequency domain, or the time-frequency domain. The principal component analysis (PCA) and k-nearest neighbor (k-NN) method were used to detect faults for bearing systems ([15]) and industrial processes [16], or the grading level of manufacturing systems [17]. The support vector machine (SVM) method was applied for fault prediction of gearboxes [18] and ship propulsion systems [19]. Zhang et al. [20] developed a fault detection system for wind turbines by using the random forest algorithm in combination with the extreme gradient boosting (XGBoost) algorithm. Abdullah [21] utilized discrete wavelet transform based artificial neural networks for fault detection of transmission lines. In [22], frequency features were extracted from vibration signals using a state-space model. A Cosine distance was designed to compare healthy frequency features and monitored features to detect early faults for a machining center. In recent years, deep learning algorithms, including CNN, RNN (Recurrent Neural Network), Stacked Autoencoders, etc., were applied for fault detection of manufacturing equipment or processes. Ince et al. [23] designed a 1-D CNN for motor fault diagnosis. Xia et al. [24] designed a CNN for fault detection for two cases, one for the motor drive system provided by the Case Western Reserve University, and another one for gearboxes. Features from vibration signals in the time and frequency domains were summarized for analysis.

Based on the above survey, it is identified that prognosis systems would be more effective in processing complex industrial situations by integrating the fog model and deep learning algorithms seamlessly. Suitable mechanisms and system structures need to be developed to improve the computation efficiency of the system.

3 System Framework and Workflow

For the system, power signals will be collected for prognosis and optimization of machining processes. According to the research of Liu et al. [25] and Sealy et al. [26], power signals from CNC machines can indicate the working and tooling conditions of machines. Power sensors are easy to deploy and the data sampling rate is much lower than that of vibration or acoustic sensors, which has been verified by experiments [33]. Meanwhile, energy consumption for machining processes can be also calculated based on power signals in order to achieve sustainable optimization of the machining system. Therefore, for this research, power signals, which are analyzed in the time domain, are used for both abnormal condition detection and multi-objective (including energy-efficient) re-scheduling optimization for detected faults.

As aforementioned analysis, the system has been designed based on the fog computing paradigm to provide a smart solution for dynamic prognosis and optimization of machining processes. The system is comprised of three layers to optimize the computation efficiency and latency of data transmission. Data transfer in the system between layers are through the MQTT protocol, taking the advantage of its lightweight communication [27], and information exchange in a dual direction simultaneously [28]. The system structure is illustrated in Fig. 1. The three layers and a coordinator between layers are briefly introduced below. Details will be elaborated in the following Sect. 4.

(1) Terminal layer: This layer is integrated with physical machine equipment via sensor devices, circuits and routers. Machining processes will follow an optimized schedule sent from the cloud layer. During entire machining processes, power signals of machines are continuously collected via power sensors and transmitted to a fog layer for further processing. When abnormal conditions of machines and tooling (e.g., severe tool wear, tool breakage, spindle failure) are detected on the fog layer, a re-schedule optimization will be triggered on the cloud layer and the generated optimized schedule will be sent to this physical layer for dynamic production adjustment.

(2) Fog layer: A trained CNN is deployed on the fog layer for efficient detection of abnormal conditions. To facilitate the efficiency and effectiveness of computing in the CNN, power signals during machining processes are first partitioned into individual stages of the machining process for each component. An algorithm of Gaussian kernel smoothness is embedded for de-noising the signals to facilitate detection. The CNN will be trained on the cloud layer and the training process will be updated for new patterns of signals when new types of components are arranged for machining in production lines. Based on the design, abnormal situations are detected using the trained CNN on fog without relatively long transmission time of the power signals to the cloud layer for judgement and decision making. Thus, signal data pertaining to operations will be kept within companies. Machines can be stopped quickly for maintenance and adjustment when potential abnormal conditions occur.

(3) Cloud layer: The cloud layer has databases to store reference signals for each machined component (called reference signals in the chapter) and typical abnormal conditions for such components. The reference signals for machined components are built based on historical data [29, 30, 33]. The CNN is re-trained when new components are arranged and reference signals are received. The trained CNN is transmitted back to the fog layer for deployment after the update. On the cloud layer, there is a multi-objective optimization algorithm to be triggered for re-scheduling when necessary.

(4) System coordinator: It is arranged between the fog and cloud layers to coordinate tasks as follows: (i) updating the knowledge database on the cloud layer for reference signals of newly machined components; (ii) re-training of the CNN using the new references; (iii) updating the trained CNN on the fog layer; (iv) triggering the scheduling optimization for the situation of abnormal situations during machining processes, and sending the optimized schedule to the terminal layer for production adjustment.

4 Detailed Design and Algorithms

4.1 Terminal Layer

The terminal layer is closely connected to machining equipment and sensor devices. Power sensors are installed in the machines to continuously monitor power. Collected power data are converted from Analog to Digital by an Analog to Digital converter (ADC). The converted raw data are transmitted to the fog layer through the MQTT protocol. When an abnormal situation identified on the fog layer, a new optimized schedule calculated on the cloud layer will be sent to the terminal layer for production adjustment accordingly.

4.2 Fog Layer

Recently, CNNs have been widely proved for high accuracy in pattern classification such as images and EEG/ECG signals [31, 32]. Considering their distinguishing characteristics, a CNN is therefore designed here for prognosis on machining processes. Furthermore, the accuracy and efficiency of the CNN will be limited if there are a lot of noises and the dataset is a bit long. In this research, pre-processing mechanisms for power signals are developed to support the CNN to process in a more effective means.

In order to better support the CNN to extract the most useful information from raw data during real-time production, monitored power signals are partitioned into individual cutting processes based on power ranges. Standby powers during cutter changes and machining breaks are lower than machining power so that power signals can be partitioned into individual cutting processes to facilitate the following CNN computing. The patterns of partitioned signals are then de-noised and smoothened based on a Gaussian kernel method to preserve the most relevant patterns.

To process monitored power signals efficiently, the power signals are partitioned into a series of separate portions according to each cutting process. A monitored power signal consists of several stages, e.g., idle, machine start-up/shut-down, machining, cutter changes, etc. The partition process is based on the power range to concentrate on the data of machining process. The partition is made based on a set of given thresholds defining the working ranges for machining. The thresholds were decided via experiments for individual machines [33]. The partition process is illustrated in Fig. 2.

The partitioned power signals will be further smoothened to de-noise the signals in order to facilitate the CNN for detecting abnormal conditions of machines and tooling effectively. An exemplar theoretical representation of power data is shown in Fig. 3a. In practical situations, features of power data collected during real production lines are hidden in fluctuated signals with noises (illustrated in Fig. 3b). The relevant patterns need to be extracted using suitable statistical methods.

For the research of pattern extraction from fluctuated signals, some statistical methods have been developed, including Principal Component Analysis (PCA) [34–36], partial least squares [37], cluster analysis [38, 39], etc. According to the research of Navi et al. [41], PCA is the most widely used approach among the above methods to handle signal data with high-dimensional (from multiple-sensor sources), noisy

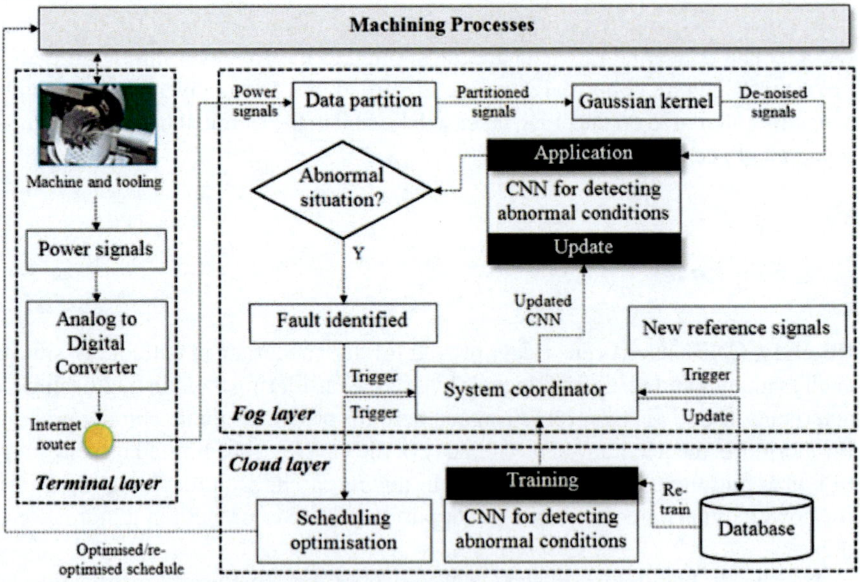

Fig. 1 Fog enabled system for prognosis and scheduling re-optimization

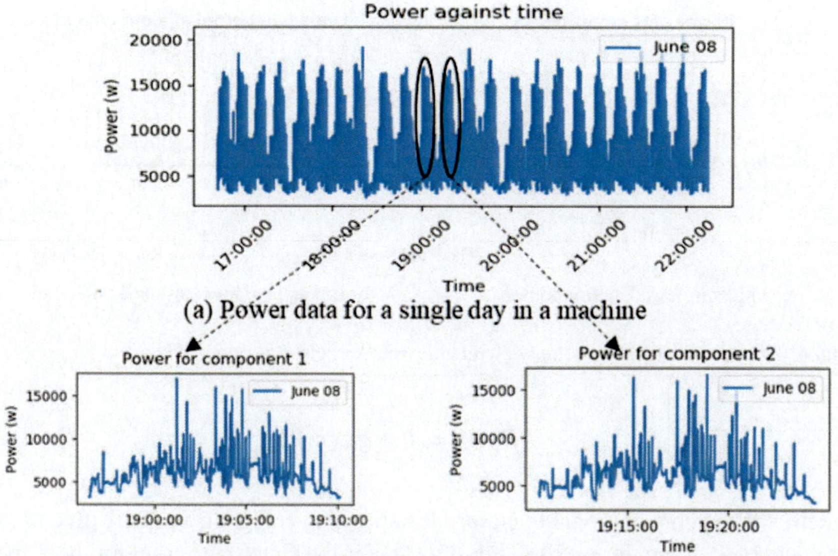

(a) Power data for a single day in a machine

(b) Power patterns for individual cutting processes partitioned from the daily data

Fig. 2 Example of partitioning power data to individual processes

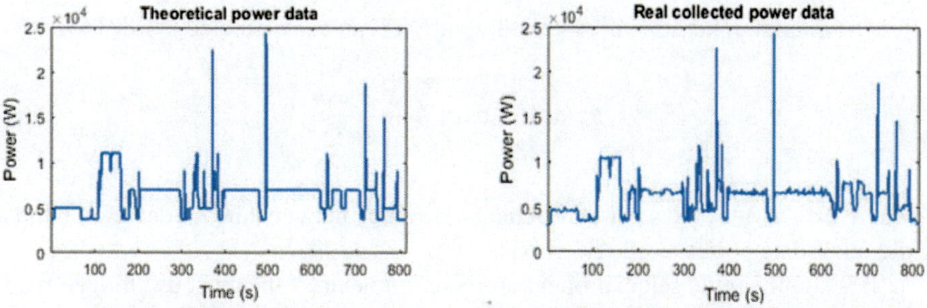

(a) A theoretical power data

(b) Power data during real production

Fig. 3 Theoretical and real power data during machining

and highly correlated features. However, PCA is not a suitable method to be applicable in this case, in which patterns are extracted from one-dimensional signal data. Furthermore, PCA is a linear method that is difficult to handle time varying dynamic and non-linear systems [40]. To handle low-dimensional data in a robust means, a Gaussian kernel model has been used and its robustness has been proved [41]. Based on the analysis, in this research, the Gaussian kernel model is chosen as a pre-processing mechanism for signal de-noising.

To support pattern extraction, data signals are smoothened by convolution with the Gaussian kernel as follows:

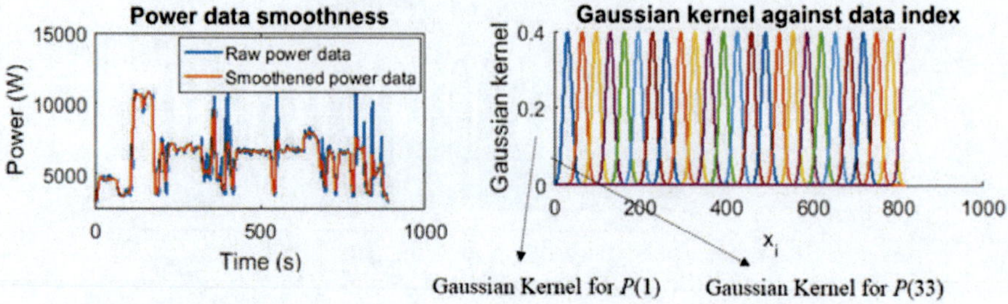

(a) Example of data smoothness (b) Example of the Gaussian kernel *P*(n)

Fig. 4 Example of signal smoothness with Gaussian kernels

$$\boldsymbol{P}_{\sigma}(\boldsymbol{i}) = \boldsymbol{P} * \boldsymbol{g}_{\sigma}(\boldsymbol{x}_{i}) \tag{1}$$

where * is the convolutional operator; $\boldsymbol{P}$ is raw power signal; $\boldsymbol{x}_i$ is the ith collected moment of $\boldsymbol{P}$ along the x-axis (Time); $\boldsymbol{g}_{\sigma}(\boldsymbol{x}_i)$ is the Gaussian kernel for the ith point with a kernel width $\boldsymbol{\sigma}$. $\boldsymbol{P}_{\sigma}(\boldsymbol{i})$ is the ith smoothened signal point calculated through this equation.

An example of power and some corresponding Gaussian kernels are shown in Fig. 4.

The Gaussian kernel with a kernel width ‘σ’ can be calculated below [36]:

$$\boldsymbol{g}_{\sigma}(\boldsymbol{x}_{i}) = \mathbf{exp}(-\frac{(\boldsymbol{x}-\boldsymbol{x}_{i})^{2}}{2\sigma^{2}}) \tag{2}$$

where $(x - x_i)^2$ is the squared Euclidean distance between the ith data point and all the other data points along the x-axis.

It is significant to select proper parameter σ to extract the most useful features for power signals. According to the research of Roueff and Vehel [42], σ can be decided below:

$$\sigma = \frac{\boldsymbol{I} - 1}{2\sqrt{\alpha \mathbf{ln}(10)}} \tag{3}$$

where I is the time support of the Gaussian kernel, which can decide the relative amplitude of the entire Gaussians. α is the attenuation coefficient and usually α = 2 [42]. σ will be investigated in Sect. 5.

To detect abnormal situations during machining process, Artificial Neural Network (ANN) has been used for fault identification and type classification. An ANN based anomaly detection algorithm was developed to detect abnormal situations by the authors [29, 33]. However, according to the previous research of the authors, the trained ANN cannot always guarantee high accuracy. Therefore, in this research, a CNN will be considered as a good alternative to detect abnormal situations, and the CNN training is processed on the cloud layer to leverage high computing powers. When a new type of component for machining is added into production, the power signal during normal machining process for the component will be collected as reference signals (usually at beginning of production, machines and cutting tools are in good conditions, and related signals are considered as reference signals for machining the components. Faulty conditions are obtained through machining life-cycles and accumulated historical data). Fault features are generated based on the reference signals according to pre-defined rules [30, 43, 44]. The CNN will be re-trained for update when there are new reference signals for new component machining. The updated CNN will be transmitted to the fog layer for efficient detection. Some abnormal situations supported by the system are described as follows and shown in Fig. 5:

Severe tool wear: the level of power shifts vertically significantly, but the level of power during the idle stage remains the same [43].

Tool breakdown: power has a sudden peak and goes back to the air cutting level [43].

Spindle failures: power has a sudden peak and an increased power level during both machining and idle stages [44].

A feature extraction method has been applied to reference signals, so that the faulty features can be amplified to achieve the higher accuracy for the CNN training. The slope and offset between every two consecutive sample points can indicate the characteristics of the curve [45]. The slope is calculated below:

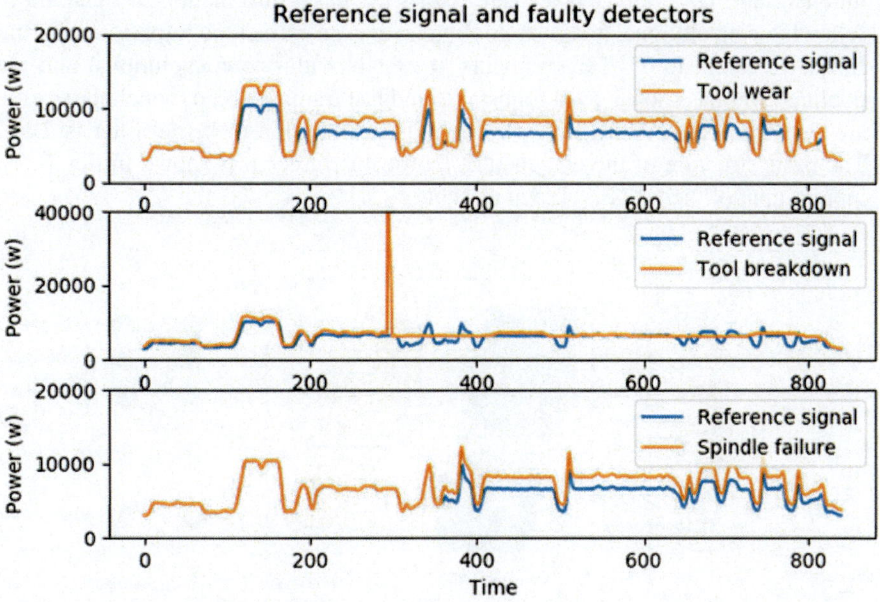

Fig. 5 Examples of reference signals and fault features

$$slope_i = \lim_{w=R} \frac{P_\sigma(i+w) - P_\sigma(i)}{w} \quad (4)$$

where $P_\sigma(i+w) - P_\sigma(i)$ is the length between two sample point under the width of w; R is the sampling rate of sensor.

$offset_i$ is the y-intercept value that each line going through two sample points. The detection index ID_i is calculated based on $slope$ and $offset$ to amplify fault features as follows:

$$ID_i = \left|(slope_i - \bar{slope})(offset_i - \bar{offset})\right| \quad (5)$$

where $\bar{slope}$ and $\bar{offset}$ are the average value of all $slope$ and$offset$, respectively. The fault detection indexes are then used as input to train the CNN. The advantage of introducing the fault detection index is benchmarked in Sect. 5.

The CNN is trained on the cloud layer based on amplified reference signals in database. The output decides whether the component is on a normal or faulty condition. A typical CNN has two combined types of layers: convolutional layers and max pooling layers. In convolutional layers, neurons are connected as rectangular grids through a kernel filter with the same weights. In the max pooling layer, rectangular grids are subsampled to extract the core information. At the last stage, all neurons are fully connected. According to research of Rajpurkar et al. [46], the first convolutional and max pooling layer can extract the basic information of data and the following convolutional and max pooling layers can extract the information further. In this research, through experiments, the CNN with one convolutional and max pooling layer has shown good robustness and fast training speed (benchmark results are shown in case studies in Section V). The dimension of kernel filter is 2 × 2. The basic structure of the designed CNN for this research is shown in Fig. 6. The input/output of the CNN is shown in Table 1.

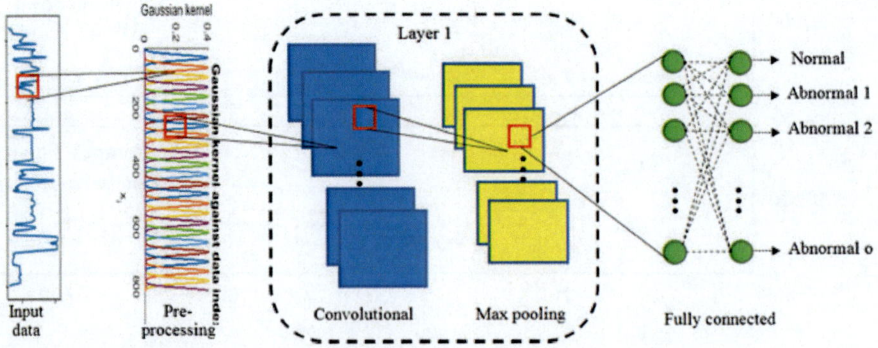

Fig. 6 Basic structure of the designed CNN

Table 1 Input and output of the CNN

Input vector (X)	Output vector ($output$)
Index 1 in power signal	Normal condition [1, 0, 0, …, 0]
Index 2 in power signal	Abnormal situation 1 [0, 1, 0, …, 0]
… …	… …
Index n in power signal	Abnormal situation o [0, 0, 0, …, 1]

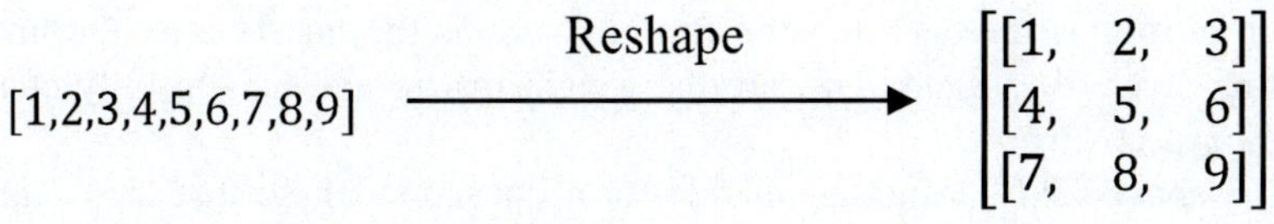

Fig. 7 An example of data reshape

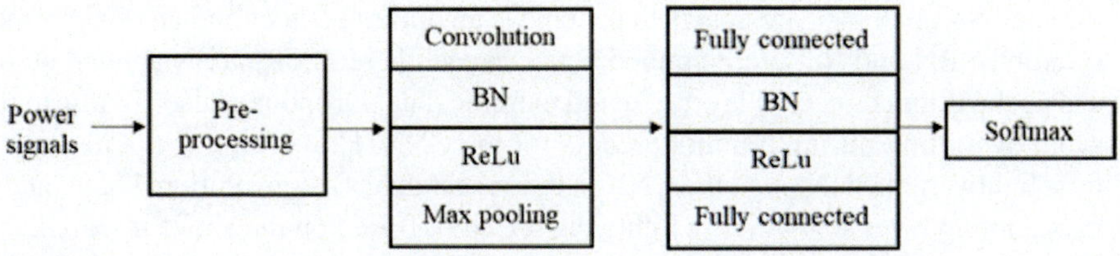

Fig. 8 BN, ReLU and Softmax introduced to enhance the CNN in this research

In order to use the CNN, the data are usually reshaped into 2-dimensional data, which is widely used in 1-dimensional signals [46]. Figure 7 explains an example of such data reshape.

In a traditional convolutional layer, data is extracted from input data through the kernel filter as follows [41]:

$$\boldsymbol{C}_{cn} = \boldsymbol{W}_{cn} * \boldsymbol{X}_j + \boldsymbol{b}_{cn} \tag{6}$$

where C_{cn} is the output matrix after convolutional layer; W_{cn} is the weight matrix of kernel filter; X_j are the input values matrix; b_{cn} is the bias value.

Max pooling follows the convolutional layer having a lower-resolution input matrix below:

$$P_{mp} = \max_{C_{cn} \epsilon S}(C_{cn}) \tag{7}$$

where P_{mp} are the outputs after the max pooling layer in the max pooling blocks; S is the max pooling block size; All the values of P_{mp} are assembled into Matrix X_{j+1} after the jth convolutional and max pooling layer, so that the size of output matrix is reduced. If there is a following convolutional and max pooling layer, X_{j+1} will be processed in the $(j+1)^{th}$ convolutional layer according to Eq. (6) until reaching the fully connected layer.

After convolutional and max pooling layers, the fully-connected layer is used in the final step for classification as calculated in the following Eq. (8).

$$output = W_{full} \times X_{final} + B_{full} \tag{8}$$

where $output$ is the classification result; X_{final} is the matrix after the final max pooling layer; W_{full} and B_{full} are the weight matrix and bias in fully connected layer, respectively.

The above CNN design is suitable for a linear model, so that the accuracy of model can be limited when dealing with non-linear problems. To improve the training efficiency, further improvements are made for the above CNN design based on recent research on CNN. During the training process, Batch Normalization (BN) [47], ReLU [48] and Softmax [49] are designed to achieve an optimal detection rate of abnormal situations. BN and ReLu are applied between each layer. Softmax is applied at the end of the fully connected layer to avoid extreme data. Dropout is also significant to avoid overfitting during training process for the CNN. However, it is not necessary for a lightweight CNN (e.g., the CNN in this research has a convolutional layer and a max pooling layer so that it is a lightweight CNN). Based on the above research, the CNN in this research is enhanced using BN, ReLU and Softmax (shown in Fig. 8). The advantage of the designed CNN in this research is validated in the case studies in Sect. 5.

When abnormal situations are identified, the machine will be temporarily disabled. The disabling time $T_{disable}$ is based on the problem complexity decided by engineers. Managed by the system coordinator, $T_{disable}$ will be sent to the terminal layer when the production schedule is re-scheduled on the cloud layer. Fog nodes usually have limited computing power and storage. Therefore, only an updated CNN is deployed on the fog layer to identify the abnormal situations without storing reference signals. Reference signals are stored in databases on the cloud layer. If new components need to be machined in the production lines, the system coordinator will be triggered and the power data signals will be collected as reference signals. As thus, the CNN will be re-trained with new reference signals on the cloud layer. The advantages of bandwidth saving and low latency under this design will be demonstrated in Section V. Meanwhile, the re-scheduling optimization algorithm requiring intensive computing will be deployed on the cloud layer and triggered by the system coordinator when necessary.

4.3 Cloud Layer

Cloud centers have the advantage of high performance of computing powers and storage spaces. Reference signals are also stored in databases on the cloud.

The scheduling optimization algorithm and CNN training generally require iterations/epochs to calculate optimum results. Therefore, they are processed on the cloud layer. At the beginning of each machining cycle, an optimization algorithm is applied to provide the optimum scheduling to minimize energy consumption, makespan and machine utilization rate by using the Fruit Fly Optimization (FFO) algorithm [33]. The optimized schedule is transmitted to the terminal layer for production control. When the cloud layer receives new reference signals, the database is updated and the CNN is re-trained with the updated reference signals, so that the CNN is able to detect and identify newly abnormal situations in the machining cycle. The updated CNN is transmitted to the fog layer for abnormal situation detection.

When any abnormal situation is identified, the machine is disabled for maintenance. The disabling time $T_{disable}(M_k)$ of Machine M_k is based on the type of abnormal situations. A multi-objective optimization mode is established for re-scheduling [33]. The total energy consumption of Machine M_k in the production line is calculated:

$$E_{total}(M_k) = E_{machine}(M_k) + E_{wait}(M_k) + E_{disable}(M_k) \tag{9}$$

where E_{total} (M_k) is the total energy consumption during all the machining phases of machine M_k; $E_{machine}$ (M_k), E_{wait} (M_k) and $E_{disable}$ (M_k) represent the energy consumption during machining, waiting and disabling phases, respectively.

The total time during all phases of all machines: $Makespan$, which is the maximum time used by all the machines in the production lines, can be calculated below:

$$Makespan = \underset{k=1}{\overset{n}{Max}}\ (T_{total}(M_k)) \tag{10}$$

The balanced utilization of production is calculated below:

$$\mu = \frac{\sum_{k=1}^{n} T_{total}(M_k)}{n} \tag{11}$$

$$Utlization\_level = \sqrt{\sum_{k=1}^{n} (T_{total}(M_k) - \mu)^2} \tag{12}$$

To optimize E_{total}, $Makespan$ and $Utlization\_level$ at the same time, Nadir and Utopia points are employed to normalize the objectives. The fitness function is calculated weighted sum of the three objectives as follows:

$$fitness = min(w_1 \cdot NE + w_2 \cdot NT + w_3 \cdot NU) \tag{13}$$

where NE, NT and NU are the normalized value of E_{total}, $Makespan$ and $Utlisation\_level$, respectively; w_1, w_2 and w_3 (weight sum = 1) are the weights for the three objectives.

The fitness is optimized by using the improved FFO algorithm to calculate a scheduler or a re-schedule from previous research [33]. The details for the optimization algorithm can be found out in [33].

4.4 Time Complexity Analysis

For the system, computation tasks are conducted on the fog layer and cloud layer, so that the time complexity for the system is analyzed for the two layers.

Fog layer.

On the fog layer, the Gaussian kernel and detection index are calculated through matrix multiplication. The time complexity can be calculated below:

$$TX_1 = O(n^3) \tag{14}$$

where TX_1 is the time complexity of the Gaussian kernel and detection index; n is the input size of the CNN.

According to Ke et al. [50], the computation process of the convolutional layer would be the most complex one in the CNN. The time complexity of the convolutional layer on the fog layer is calculated below [51]:

$$TX_2 = O(\sum_{l=1}^{d} n_{l-1} \times {s_l}^2 \times n_l \times {m_l}^2) \tag{15}$$

where TX_2 is the time complexity of the convolutional layer on the fog layer; l is no. of a convolutional layer in the CNN; d is the total number of convolutional layers; n_l is the number of filters in the lth layer; n_{l-1} is also known as the number of input channels of the lth layer;s_l is the length of the filter; m_l is the spatial size of the output feature map.

Cloud layer.

On the cloud layer, both the optimization algorithm and CNN training are performed. The time complexity of optimization can be calculated below [54]:

$$TX_3 = \mathrm{O}(n_g \times n_O \times {n_p}^3) \tag{16}$$

where TX_3 is the time complexity of optimization; n_g is the number of generations in the algorithm; n_0 is the number of optimization objectives; n_p is the population size.

According to the research of Bianchini and Scarselli [55], the maximum time complex for training a CNN is:

$$TX_4 = O(4^{h^2}(nh)^{n+2h}) \tag{17}$$

where TX_4 is the time complexity of the CNN; h is the total number of hidden neurons in all the layers; n is the number of the input signals.

5 System Deployment and Analysis

5.1 System Setup and Deployment

For the system, on the terminal layer, power sensors are mounted with machining equipment. Power data are collected by the Wemos Wi-Fi board (https://www.wemos.cc/), which is a cost-effective platform with an Analog to Digital converter and an embedded Wi-Fi module. Monitored power data are transmitted to the fog layer via Wi-Fi. The fog layer consists of an Internet router and Raspberry Pi (https://www.raspberrypi.org/), Raspberry Pi is located right beside the Wemos Wi-Fi board to receive data within the shop floor. Received data on the fog layer are processed by the trained CNN deployed into the Raspberry Pi. Reference signals are stored on the cloud layer to train the CNN. Managed by the system coordinator, the CNN is re-trained when necessary, and the trained CNN is transmitted back to the fog layer for re-deployment. Figure 9 illustrates the system and some machining centers for this deployment. Table 2 shows the specification of the Raspberry Pi used on the fog layer and the cloud server.

The fog enabled prognosis system was deployed in a UK company for this industrial trial. The company specializes in precision machining for automotive, aeronautical and nuclear components. Two CNC machine centers (MX520, MAZAK VTC-800/20SR) were monitored for approximately three months. Raw materials towers and loading/unloading robotic arms were integrated into the production line to implement automation processes. The production is automatic, working for 6 days per week (from Monday to Saturday). Operators/maintenance engineers were scheduled to check the status of machining processes.

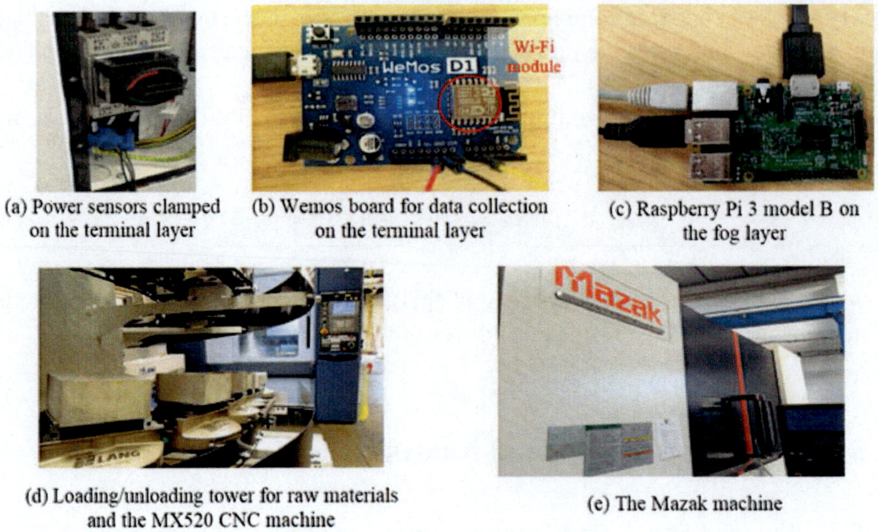

(a) Power sensors clamped on the terminal layer

(b) Wemos board for data collection on the terminal layer

(c) Raspberry Pi 3 model B on the fog layer

(d) Loading/unloading tower for raw materials and the MX520 CNC machine

(e) The Mazak machine

Fig. 9 Deployment of the fog computing system into machining processes

Table 2 Specification of the Raspberry Pi and cloud server

Hardware	Connectivity	Processor	Memory	Storage
Raspberry Pi	2.4 GHz/5 GHz IEEE 802.11	1.4 GHz	1 GB	SD card (16 GB)
Cloud server	2.4 GHz/5 GHz IEEE 802.11	4.2Ghz	32 GB	1 TB

5.2 *Analysis on Gaussian Kernel Modelling and Fault Index*

In the system, the Gaussian kernel is applied to monitored raw power signals according to Eqs. (1–3). Based on this pre-processing mechanism, the identification process of the most useful features from machining processes by the CNN will be much quicker and more accurate. Via experiments on a number of samples and statistical analysis, when $\sigma = 10$, the smoothened signal can keep the most the machining features. Figure 10 illustrates the smoothness with different values of σ. As shown in Fig. 11, an example of fault detection indexes is created according to Eqs. (4)-(5). As shown in Table 3, the Mean Absolute Percentage of Error (MAPE) between a faulty signal and a reference signal is calculated for both smoothened data and fault detection index. It can clearly indicate that the faulty features can be significantly magnified to facilitate feature identification by the CNN process.

In order to show the effectiveness of introducing the Gaussian kernel and fault detection index to improve the training process of the CNN, both signal data with/without smoothness and index are utilized for comparison. The benchmark results are shown in Fig. 12. The training process of the CNN with smoothened and indexed data achieved 98% accuracy in 12 epochs. The training process only with indexed data achieved 98% accuracy in 21 epochs. Other training processes were not

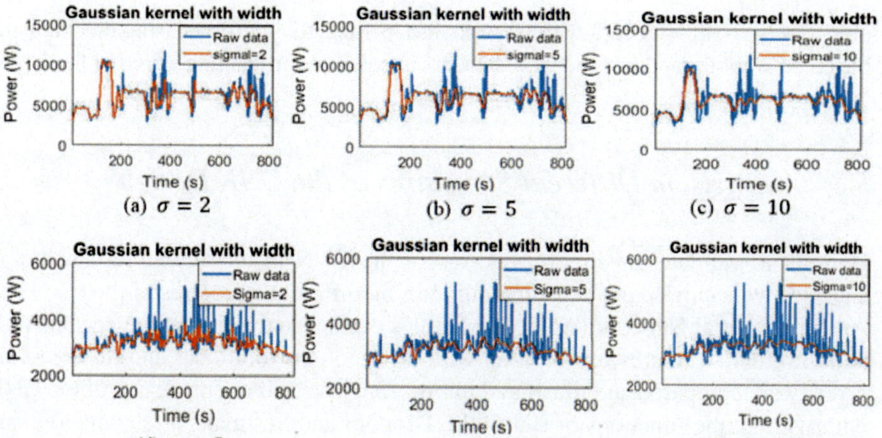

(a) $\sigma = 2$ (b) $\sigma = 5$ (c) $\sigma = 10$

(d) $\sigma = 2$ (e) $\sigma = 5$ (f) $\sigma = 10$

Fig. 10 Smoothened signals by the Gaussian kernel with different width

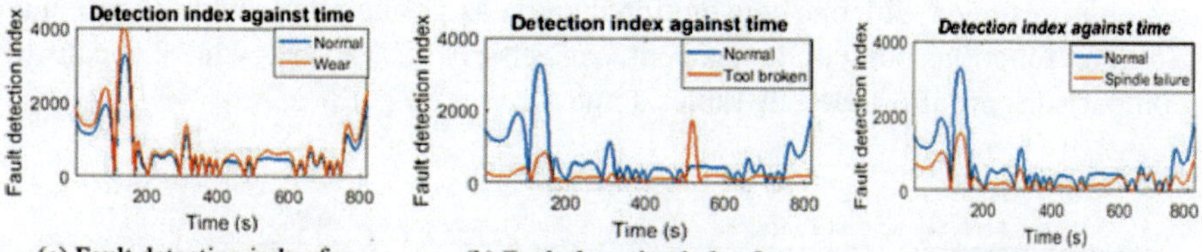

(a) Fault detection index for tool wear (b) Fault detection index for tool breakage (c) Fault detection index for spindle failure

Fig. 11 Fault detection index

Table 3 MAPEs for smoothened data and fault detection index

	Tool wear	Tool breakage	Spindle failure
Smoothened data	83.66	86.75	23.68
Fault detection index	260.23	699.99	522.28
Improvement in magnification	211.06%	706.90%	2105.57%

Fig. 12 Benchmarks on the CNN training with processed and raw signal data

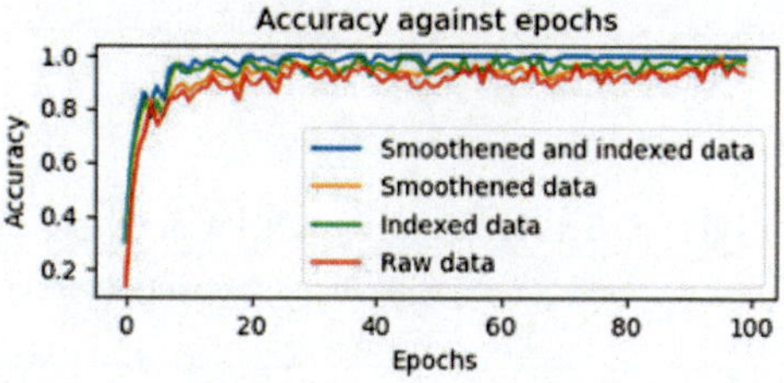

able to achieve 98% accuracy in 100 epochs. As thus, the pre-processing mechanisms based on smoothened and indexed data can accelerate the training speed of the CNN.

5.3 Analysis on Different Structures of the CNN Design

In order to validate the structure design of the CNN in this research, benchmark analyses were carried out by executing training different structures of the CNN and Artificial Neural Network (ANN) for 5 times each. The training processes on ANN and different structures of the CNN with different convolutional and max pooling layers were compared and illustrated in Fig. 13a. The training processes of the CNN with/without the functions of BN, Relu, Dropout and Softmax were compared and illustrated in Fig. 13b. In this case study, the accuracy of ANN (92.91%) is much lower than that of CNN (98%). Therefore, ANN cannot satisfy the requirement of real production with low fault-tolerance. It can be observed clearly that the CNN, which is designed with one convolutional and max pooling layer with BN, Relu and Softmax functions but without dropout, can achieve 98% accuracy in 12 epochs (the comparisons are illustrated in Tables 4 and 5).

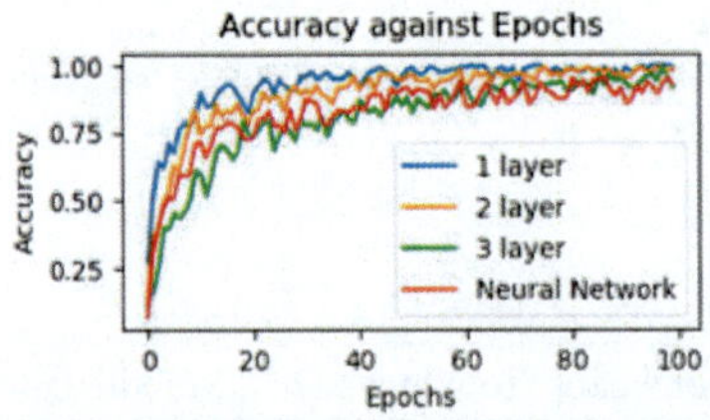

(a) Benchmark of Neural Network and CNN with different convolutional and max pooling layers

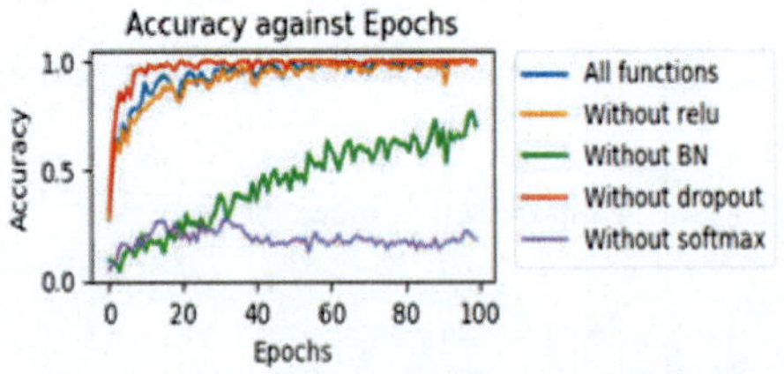

(b) Benchmark of CNN with different functions

Fig. 13 Comparison on training the CNN with different design

Table 4 Benchmark on Training of the ANN and CNN with different layers

	One layer	Two layers	Three layers	ANN
Epochs to achieve 98%	12	60	98	Did not achieve

Table 5 Benchmark on the CNN with various functions

	with all functions	without Relu	without BN	Without dropout	without Softmax
Epochs to achieve 98%	44	50	Did not achieve	12	Did not achieve

5.4 Analysis on System Performance Improvement

The system separates the computation of the training and execution processes of the CNN into the fog and cloud layers respectively. This "divide and conquer" approach developed in this chapter can significantly optimize the data traffic and computation efficiency in comparison with cloud solutions.

During this industrial trial, the total quantity of collected power data were 12.98 GB. Owing to the fog design, only 3.36 GB data were transmitted to the cloud layer. As thus, 70.26% bandwidth was saved in comparison with a cloud solution [33]. For 1 Mb data, it roughly took 1.31 s for monitoring signals to be transmitted to and processed on fog nodes, while it roughly took 3.96 s for data transmitted and processed in a cloud center (it was also observed that data transmission to the cloud solution was even much slower when there are transmission congestions in the Internet). The improvements of the time spent for data transfer in the fog system was 47.02%. The summary of the system performance is shown in Table 6.

The fog design presented in this research was further justified through the following computing and storage. Based on the configuration of Table 2, calculation time and memory usage for all the algorithmic elements in the industrial trials are shown in Table 7. It indicates that pre-processing and CNN for prognosis can be effectively implemented on the fog layer based on the capacity of the edge devices. However, the scheduling optimization and CNN training demonstrated high time complexity and should be arranged on the cloud. Thus, this "divide and conquer" fog architecture can ensure close interaction efficiency between prognosis and working machines, and leverage the capacity of cloud servers for intensive computation.

Table 6 Comparison of the fog system and a cloud system [33]

Method/Improvement	Total monitored data (GB)	Transmitted data to cloud (GB)
This fog based system	12.98	3.86
Cloud based system [33]	12.98	12.98
Bandwidth saving	70.26%	
Reduction of time spent for data transfer	47.02%	

Table 7 Real calculation time and memory usage for complexity analysis

Fog layer	Gaussian kernel	Detection index	CNN for prognosis	Total
Computing time (s)	0.21	0.12	0.98	1.31
Memory usage (MB)	1.37	1.10	3.76	6.32
Cloud layer	Optimization		CNN training	Total
Computing time (s)	3.18		1253.91	1257.09
Memory usage (MB)	20.12		291.93	312.05

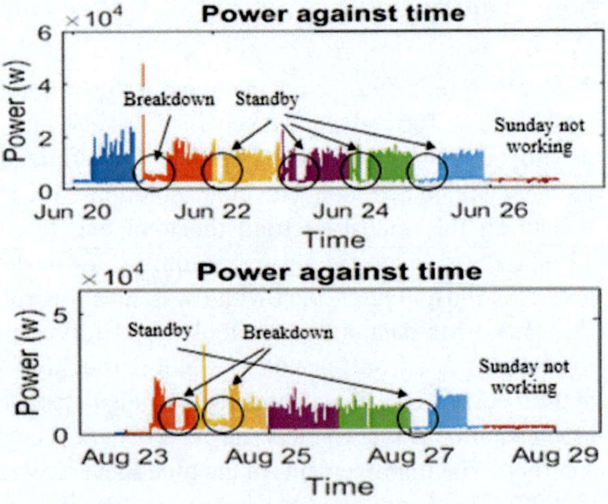

Fig. 14 Production with a number of breakdown in two weeks as examples

5.5 *Analysis on Improved Efficiency of Machining System*

For the company, before the deployment of the fog system, the production experienced various breakdown due to the issues of tooling, machines and long-time standby for manual check on machining conditions (examples before the deployment of the system are illustrated in Fig. 14). After the deployment of the fog enabled prognosis system, the efficiency of energy and production was improved approximately for 29.25% and 16.50% on average.

6 Conclusions

This chapter presents a fog and CNN enabled system for dynamic prognosis and optimization of machining processes to address practical manufacturing situations. The functions and innovations include:

(1) A three-layer fog and cloud model are developed to optimize computing resource utilization and minimize data traffic. Enabled by the fog computing model, the system can overpass the prevalent low-efficiency issue of cloud based systems rooted from the high latency of data transfer in industrial networks. Complex analysis and computation in the system, like algorithm training, data analysis, scheduling/re-scheduling optimization, are well balanced with the design.

(2) Modern machining companies especially SMEs are characterized by low-volume and high-mix customized production. The occurrence of faults during customized machining processes is uncertain. To address complex conditions effectively, the system is augmented with a CNN to dynamically mine sensitive fault features

from large-scale signals acquired in practical production. Pre-processing mechanisms are designed for effective signal de-noising to enhance the performance of the CNN in practical manufacturing situations. Benchmark experiments and analyses are conducted to determine suitable structures and parameters for the CNN design.

(3) The performance of the system was assessed in a European machining company and achieved satisfactory results. With significant improvements on bandwidth reduction and system efficiency in case studies, the effectiveness and applicability of this system to manufacturing practices were clearly indicated. The system is applicable to other manufacturing processes by training the CNN using fault features from those processes. The research provides manufacturing practitioners an innovative solution based on the state-of-the-art industrial artificial intelligence.

Currently, the training efficiency is still limited in re-training the CNN when new component patterns are introduced. The transfer learning technology can resolve the issue by training the CNN based on the parameters of the trained CNN. This topic will be researched in the future work.

References:

1. Lee J, Wu F, Zhao W, Ghaffari M, Liao L, Siegel D (2014) Prognostics and health management design for rotary machinery systems - Reviews, methodology and applications. Mech Syst Signal Process 42(1–2):314–334
2. Gao R, Wang LH, Teti R, Dornfeld D, Kumara S, Mori M, Helu M (2015) Cloud XE "Cloud" -enabled prognosis for manufacturing. CIRP Ann 64(2):749–772
3. Tao F, Qi Q, Liu A, Kusiak A (2018) Data-driven smart manufacturing. J Manufact Syst 48:157–169
4. Baccarelli E, Naranjo P, Scarpiniti M, Shojafar M, Abawajy J (2017) Fog XE "Fog" of everything: Energy-efficient networked computing architectures, research challenges, and a case study. IEEE Access 5:9882–9910
5. Hu P, Dhelim S, Ning H, Qiu T (2017) Survey on fog computing: Architecture, key technologies, applications and open issues. J Network Comput Appl 98:27–42
6. Mukherjee M, Shu L, Wang D (2018) Survey of fog computing: Fundamental, network applications, and research challenges. IEEE Communications Surveys & Tutorials. pp 1–1
7. Wu D, Liu S, Zhang L, Terpenny J, Gao R, Kurfess T, Guzzo J (2017) A fog computing-based framework for process monitoring and prognosis in cyber-manufacturing. J Manufact Syst 43:25–34
8. Lin C, Yang J (2018) Cost-efficient deployment of fog computing systems at logistics centres in Industry 4.0. IEEE Trans Ind Inform 14(10):4603–4611
9. Wan J, Chen B, Wang S, Xia M, Li D, Liu C (2018) Fog XE "Fog" computing XE "Fog computing" for energy-aware load balancing and scheduling in smart factory. IEEE Trans Industr Inf 14(10):4548–4556
10. O'Donovan P, Gallagher C, Bruton K, O'Sullivan D (2018) A fog computing industrial cyber-physical system for embedded low-latency machine learning Industry 4.0 applications. Manufact Lett 15:139–142
11. Mohamed N, Al-Jaroodi J, Jawhar I (2018) Utilizing fog computing for multi-robot systems. 2018 Second IEEE International Conference on Robotic Computing (IRC), CA, USA, January 31–February 2
12. Liu R, Yang B, Zio E, Chen X (2018) Artificial intelligence for fault diagnosis of rotating machinery: A review. Mech Syst Signal Process 108:33–47

13. Wang J, Ma Y, Zhang L, Gao RX, Wu D (2018) Deep learning for smart manufacturing: Methods and applications. J Manufact Syst 48:144–156
14. Zhao R, Yan R, Chen Z, Mao K, Wang P, Gao RX (2018) Deep learning XE "Deep learning" and its applications to machine health monitoring. Mech Syst Signal Process 115:213–237
15. Tian J, Morillo C, Azarian MH, Pecht M (2016) Motor bearing fault detection using spectral kurtosis-based feature extraction coupled with k-nearest neighbor distance analysis. IEEE Trans Industr Electron 63(3):1793–1803
16. Zhou Z, Wen C, Yang C (2016) Fault isolation based on k-nearest neighbor rule for industrial processes. IEEE Trans Industr Electron 63(4):2578–2586
17. Li F, Wang J, Tang B, Tian D (2014) Life grade recognition method based on supervised uncorrelated orthogonal locality preserving projection and k-nearest neighbor classifier. Neurocomputing 138:271–282
18. Li C, Sanchez R-V, Zurita G, Cerrada M, Cabrera D, Vasquez RE (2015) Multimodal deep support vector classification with homologous features and its application to gearbox fault diagnosis. Neurocomputing 168:119–127
19. Zhou J, Yang Y, Ding S, Zi Y, Wei M (2018) A fault detection and health monitoring scheme for ship propulsion systems using SVM technique. IEEE Access 6:16207–16215
20. Zhang D, Qian L, Mao B, Huang C, Huang B, Si Y (2018) A data-driven design for fault detection of wind turbines using random forests and XGboost. IEEE Access 6:21020–21031
21. Abdullah A (2018) Ultrafast transmission line fault detection using a DWT-based ANN XE "ANN" . IEEE Trans Ind Appl 54(2):1182–1193
22. Luo B, Wang H, Liu H, Li B, Peng F (2019) Early fault detection of machine tools based on deep learning and dynamic identification. IEEE Trans Industr Electron 66(1):509–518
23. Ince T, Kiranyaz S, Eren L, Askar M, Gabbouj M (2016) Real-time motor fault detection by 1-D convolutional neural networks. IEEE Trans Industr Electron 63(11):7067–7075
24. Xia M, Li T, Xu L, Liu L, de Silva CW (2018) Fault diagnosis for rotating machinery using multiple sensor and convolutional neural networks. IEEE/ASME Trans Mechatron 23(1):101–110
25. Liu Z, Guo Y, Sealy M, Liu Z (2016) Energy consumption and process sustainability of hard milling with tool wear progression. J Mater Process Technol 229:305–312
26. Sealy M, Liu Z, Zhang D, Guo Y, Liu Z (2016) Energy consumption and modeling in precision hard milling. J Clean Product 135:1591–1601
27. Chooruang K, Mangkalakeeree P (2016) Wireless heart rate monitoring system using MQTT. Procedia Comput Sci 86:160–163
28. Schmitt A, Carlier F, Renault V (2018) Dynamic bridge generation for IoT data exchange via the MQTT protocol. Procedia Comput Sci 130:90–97
29. Liang YC, Li WD, Wang S, Lu X (2019) Big Data based dynamic scheduling optimization for energy efficient machining. Engineering 5:646–652
30. Wang S, Liang YC, Li WD, Cai X (2018) Big data enabled intelligent immune system for energy efficient manufacturing management. J Clean Product 195:507–520
31. Franc V, Čech J (2018) Learning CNN XE "CNN" s XE "CNNs" from weakly annotated facial images. Image Vis Comput 77:10–20
32. Banerjee S, Das S (2018) Mutual variation of information on transfer-CNN XE "CNN" for face recognition with degraded probe samples. Neurocomputing 310:299–315
33. Liang YC, Lu X, Li WD, Wang S (2018) Cyber Physical System and Big Data enabled energy efficient machining optimization. J Clean Product 187:46–62
34. Liu Q, Qin S, Chai T (2013) Decentralized fault diagnosis of continuous annealing processes based on multilevel PCA XE " principle component analysis (PCA) ". IEEE Trans Autom Sci Eng 10(3):687–698
35. Guo Y, Li G, Chen H, Hu Y, Li H, Xing L, Hu W (2017) An enhanced PCA XE " principle component analysis (PCA) " method with Savitzky-Golay method for VRF system sensor fault detection and diagnosis. Energy Build 142:167–178
36. Kyprianou A, Phinikarides A, Makrides G, Georghiou G (2015) Definition and computation of the degradation rates of photovoltaic systems of different technologies with robust Principal Component Analysis. IEEE J Photo 5(6):1698–1705

37. Wu J., Chen W., Huang K., Tan T., 2011. Partial least squares based subwindow search for pedestrian detection. 2011 18th IEEE International Conference on Image Processing.
38. Yu Z, Chen H, You J, Liu J, Wong H, Han G, Li L (2015) Adaptive fuzzy consensus clustering framework for clustering analysis of cancer data. IEEE/ACM Trans Comput Biol Bioinf 12(4):887–901
39. Yan R, Ma Z, Kokogiannakis G, Zhao Y (2016) A sensor fault detection strategy for air handling units using cluster analysis. Autom Construct 70:77–88
40. Navi M, Meskin N, Davoodi M (2018) Sensor fault detection and isolation of an industrial gas turbine using partial adaptive KPCA XE "principle component analysis (PCA)". J Process Control 64:37–48
41. Rimpault X, Bitar-Nehme E, Balazinski M, Mayer J (2018) Online monitoring and failure detection of capacitive displacement sensor in a Capball device using fractal analysis. Measurement 118:23–28
42. Roueff F, Vehel J (2018) A regularization approach to fractional dimension estimation. World Scientific Publisher, Fractals and Beyond
43. Teti R, Jemielniak K, O'Donnell G, Dornfeld D (2010) Advanced monitoring of machining operations. CIRP Annals – Manufact Technol 59:717–739
44. Axinte D, Gindy N (2004) Assessment of the effectiveness of a spindle power signal for tool condition monitoring in machining processes. Int J Prod Res 42(13):2679–2691
45. Feng Z, Zuo M, Chu F (2010) Application of regularization dimension to gear damage assessment. Mech Syst Signal Process 24(4):1081–1098
46. Rajpurkar P, Hannun A, Haghpanahi M, Bourn C, Ng A (2018) Cardiologist-level arrhythmia detection with convolutional neural networks. arXiv:1707.01836
47. Ioffe S, Szegedy C (2015) Batch normalization: Accelerating deep network training by reducing internal covariate shift. arXiv:1502.03167
48. Priddy K, Keller P (2005) Artificial neural networks. Bellingham, Wash. <1000 20th St. Bellingham WA 98225–6705 USA>: SPIE
49. Hinton G, Srivastava N, Krizhevsky A, Sutskever I, Salakhutdinov R (2012) Improving neural networks by preventing co-adaptation of feature detectors. arXiv:1207.0580v1
50. Ke X, Cao W, Lv F (2017) Relationship between complexity and precision of convolutional neural networks. Proceedings of the 2017 2nd International Symposium on Advances in Electrical, Electronics and Computer Engineering (ISAEECE 2017)
51. He K, Sun J (2014) Convolutional neural networks at constrained time cost. arXiv:1412.1710

Big Data Enabled Intelligent Immune System for Energy Efficient Manufacturing Management

S. Wang, Y. C. Liang, and W. D. Li

Abbreviations

$E_{\text{waiting}}(M_i)$	The energy consumption of Machine M_i during waiting.
$E_{machining}(M_i, J_j)$	The energy consumption of machining Component J_j by machine M_i.
$E_{machining}(M_i)$	The energy consumption of Machine M_i during machining.
$E_{total}(M_i)$	The energy consumed during all the phases of Machine M_i.
F_1	A measure of a test's accuracy.
FN	False Negative.
FP	False Positive.
M	The total number of components to be machined.
$MAPE$	Mean Absolute Percentage Error.
$MAPE_L$, $MAPE_U$	Lower bound and upper bound of Mean Absolute Percentage Error.
ME	The Mean Error between two patterns.
N	The bigger number of the samples of the two patterns.
N_1	The total number of the measured power points in $S_{measure}$.
P_L, P_U	Lower bound and upper bound of a standard energy pattern.
$Precision$	The proportion of all the positive predictions that are correct.
$Recall$	Proportion of all the real positive observations that are correct.
$S_{measure}$	Measured energy pattern for machining a component.
$S_{standard}$	Standard energy pattern for machining a component.
$S'_{measure}$	The shifted $\text{S}_{\text{measure}}$ to $\text{S}_{\text{standard}}$.

S. Wang · Y. C. Liang · W. D. Li (✉)
Faculty of Engineering, Environment and Computing, Coventry University, Coventry, UK
e-mail: weidong.li@coventry.ac.uk

W. D. Li
School of Logistics Engineering, Wuhan University of Technology, Wuhan, China

W. Li et al. (eds.), *Data Driven Smart Manufacturing Technologies and Applications*, Springer Series in Advanced Manufacturing,
https://doi.org/10.1007/978-3-030-66849-5_3

t_{21}	Time delay.
TP	True Positive.
X_{coef}	Cross-correction coefficient.
z_i^N	The upper bound of the ith objective function.
z_i^U	The lower bound of the ith objective function.
$\mu_{measure}$, $\mu_{standard}$	The means of time patterns.
$\sigma_{Smeasure,Sstandard}$	The cross-covariance between the pair of patterns.

1 Introduction

Ambitious goals to achieve significant energy savings in manufacturing have been widely set by major economies such as Europe, China and USA [1]. Recent research has summarized that the energy efficiency indicators of manufacturing on a national or sectional level have been defined, but relevant sustainable process management solutions for companies have not been effectively implemented. There is a strong need to foster the relevant empirical applied research in manufacturing companies [2].

In a manufacturing shop floor, dynamics during manufacturing execution lifecycles are major challenges preventing companies from maintaining the best energy performance as well as manufacturing quality and productivity. The current management systems like process planning, scheduling and execution systems used in European factories are based on relatively static manufacturing information as a cascading decision-making process [3]. On the other hand, dynamics could be from various aspects of manufacturing execution lifecycles, such as frequent job order or priority changes, ambient working conditions, and unexpected delays. Manufacturing systems and tooling are also prone to aging and degrading, resulting in dimensional and geometric deviations of manufactured components. The dynamics generate unexpected breaks and unnecessary inspection, standby, repairing and maintenance of manufacturing systems, leading to time, energy and resource waste [4]. To address dynamics in manufacturing, smart sensors, Cyber Physical System (CPS) and Big Data analytics have been increasingly deployed in industrial shop floors to support condition monitoring and diagnosis [5]. However, due to the large quantities of monitored data and diverse manufacturing models, relevant data collection and analysis to support energy efficient manufacturing are still inefficient and error-prone.

In this chapter, an innovative Big Data enabled Intelligent Immune System (I^2S) has been designed to achieve energy efficient manufacturing optimization via energy monitoring, analysis and manufacturing re-scheduling. Different from conventional management approaches that are based on pre-defined manufacturing conditions, I^2S is enabled by CPS to collect energy (electricity) consumption data of manufacturing processes so as to monitor the dynamics and condition changes of the manufacturing systems efficiently. Artificial Neural Networks (ANNs)-based algorithms and statistical analytics tools are used to identify the energy consumption patterns of manufactured components. An artificial immune mechanism is then applied to counter significantly varying conditions during the lifecycle of the manufacturing process. A re-scheduling adjustment is triggered if necessary thereby achieving energy savings and maintaining optimized performance (productivity and balanced level of machine utilization) for an entire manufacturing execution lifecycle. In this research, Computer Numerical Controlled (CNC) machining processes have been used for system development, validation and industrial deployment.

The research innovations and characteristics of I^2S are below:

The CPS, ANNs, immune mechanism and re-scheduling optimization algorithm have been effectively integrated as innovative manufacturing intelligence to support energy efficient optimization over manufacturing execution lifecycles. The immune mechanism has been designed to address abnormal working conditions in a systematic means;

The collected energy data from machining systems are stored as "Big Data", owing to the long time and high frequency of electricity data collection (i.e., 3-V features for Big Data—high Volume of collected data, high Velocity of collecting data, and high Variety of data patterns generated from different machined components). Deep learning and Convolutional Neural Networks (CNNs) are the state-of-the-art artificial intelligent technologies for Big Data processing [6]. However, their disadvantages are the requirements of long time, high computational power (GPU) and large training sets [7]. To meet industrial requirements, in this research, a "divide and conquer" strategy has been designed to improve the efficiency and robustness of Big Data analysis. That is, the energy Big Data has been pre-processed and partitioned, and three-layer ANNs and statistical analysis tools have been introduced to learn and distinguish patterns efficiently so as to support abnormal condition processing;

I^2S has been validated through real-world industrial deployment into some European machining companies located in Sweden, U.K. and Spain for over six months respectively to demonstrate the significant potentials of the system's applicability in practice. Significant sustainability improvements on the environmental, economic and social aspects have been achieved by adopting I^2S into the European factories (less unexpected breaks and scheduling optimization to improve energy efficiency and productivity, intelligent monitoring and prognosis to 6avoid tedious human intervention and errors).

2 Literature Survey

Scheduling is a critical decision-making stage in manufacturing to minimize lifecycle cost, enhance adaptability to manufacturing and improve manufacturing sustainability. In the past decades, scheduling optimization has been widely researched (comprehensive surveys have been made in [8, 9]). In the research, optimization objectives are from the aspects of lead-time, makespan and cost minimization, and/or the most balanced utilization level of machines. In recent years, in line with the trend on achieving sustainability in manufacturing management, energy efficiency has been increasingly considered as an important optimization objective. Based on various developed energy models, scheduling optimization algorithms have been developed and applied to improve the energy efficiency of manufacturing processes [4]. An improved Particle Swarm Optimization (PSO) approach was designed to address dynamic scheduling under unexpected disruptions to reduce energy consumption and makespan simultaneously [10]. A new multi-objective Genetic Algorithm (GA) and NSGA-II algorithm was developed to minimize the total non-processing electricity consumption and total weighted tardiness [11]. In the algorithm, a function for parent and children combination and elitism to improve optimization further was developed. In the work [12], based on a multi-level energy model and grey relational analysis to optimize machining parameters, a GA was developed to optimize the makespan and energy consumption. Based on real-time monitored data, an enhanced Pareto-based bee algorithm was designed to optimize energy consumption and productivity [13]. Salido et al. [14] developed a memetic algorithm to minimize energy consumption under makespan constraints. However, the above research is based on relatively static machining resource information (e.g., the optimization algorithms are based on prior experimental results before manufacturing execution) so that dynamics during manufacturing execution lifecycle are unable to be considered effectively (e.g., machining resources and working conditions are assumed unchanged though there are various ambient elements and job dynamics during the execution lifecycle of customized production). To overcome the limit, Cai et al. [15] designed an intelligent immune algorithm that is analogous to the biological immune mechanism to eliminate disturbances produced during manufacturing scheduling operations. In the research, a framework to map an intelligent manufacturing process into an artificial immune system was developed. Based on the biological immune system that potentially offers interesting features to face the threats (bacteria, viruses, cancers, etc.), Darmoul et al. [16] investigated the application of intelligent immune algorithms to monitor and control manufacturing systems at the occurrence of disruptions. However, the above immune works are not employed to practical applications yet. Meanwhile, the reported immune systems are based on a Non-self/Self (N/S) mechanism, which considers non-self elements as problems. This mechanism is not flexible enough to effectively process various dynamics of manufacturing processes (analysis is given in Sect. 2 Chap. 4). A summary of the above related research is given in Table 1.

Another trend is to use in-process monitoring data and Big Data analytics for manufacturing optimization such as scheduling and condition-based maintenance to achieve manufacturing lifecycle optimization. A new scheduling system for selecting dispatching rules in real time by combining the techniques of simulation, data mining, and statistical process control charts was developed [17]. In the research, the developed scheduling system extracts knowledge from data coming from manufacturing environments. The knowledge provides the system the adaptiveness to changing manufacturing environment and enhanced adaptability and quality of its decisions. Data analytics were also developed for condition-based maintenance and real-time manufacturing optimization. A data mining technique to conduct logical analysis of data for condition-based maintenance was proposed in [18]. Kusiak and Verma [19] established ANNs to identify bearing faults in wind turbines based on real-time data.

Table 1 Recent research for sustainable and lifecycle manufacturing

Works	Input	Optimisation targets	Research methods
Metan et al. (2010) [17]	Data coming from the manufacturing environment	Average tardiness minimization	Process control charts to monitor and update the performance of the decision tree
Fang et al. (2011) [21]	Machining width, feed per tooth, machining speed and specific machining energy	Makespan, peak power demand, and carbon footprint	Empirical models and case studies of machining cast iron plates with slots
He et al. (2012) [22]	CNC codes	Energy consumption for spindle, axis feed, tool changes, coolant pump and fixed energy consuming units	Empirical models for spindle, axis feed, tool changes, coolant pump and fixed energy units
Yan and Li (2013) [24]	Material removal rate, idle power, machine tool specific coefficients and standby power	Energy consumption model	Thermal equilibrium and empirical
Winter et al. (2014) [23]	Machining depth, machining speed and dressing speed	Energy consumption	Sensitivity analysis method
Purarjomandlangrudi et al. (2014) [20]	Monitoring data for rolling-element bearing failures	Minimisation of failure rate	Kurtosis and Non-Gaussianity Score (NGS)
Wang et al. (2015) [4]	Spindle speed, machining speed, depth of cut and width of cut Number of machines and the number of jobs to be processed	Surface quality, energy consumption and machining removal rate Energy consumption for idle, working, tool change and set-up	ANNs to establish a model for surface quality and energy consumption Empirical models for idle, working, tool change and set-up
Yan et al. (2016) [12]	Material removal rate, spindle speed Number of machines and the number of jobs to be processed	Idle power and operation power, energy consumption for processing, set-up, transportation, standby, and overhead	Off-line experiments Empirical models for processing, set-up, transportation, standby and overhead

(continued)

Table 1 (continued)

Works	Input	Optimisation targets	Research methods
Tang et al. (2016) [10]	Manufacturing components, sequences, manufacturing resources	Schedule, process planning	Non-Self/Self immune system, Artificial Neural Networks
Darmoul et al. (2016) [16]	Material unavailability, resource failures, unavailability of operators, rush orders	Monitoring and control of manufacturing systems at the occurrence of disruptions	Non-Self/Self immune system framework

Table 2 Input and output of the component energy classification-ANNs

Input vector	Output vector
Energy input point *1*	Component category *1* [*1*, *0*, *0*, ..., *0*]
Energy input point *2*	Component category *2* [*0*, *1*, *0*, ..., *0*]
Energy input point *3*	Component category *3* [*0*, *0*, *1*, ..., *0*]
... ...	... ...
Energy input point *n*	Component category *o* [*0*, *0*, *0*, ..., *1*]

An anomaly detection-based data mining approach was developed to discriminate defect examples of rolling-element bearing failures [20]. Two features, i.e., kurtosis and Non-Gaussianity Score (NGS), are extracted to develop anomaly detection algorithms. Nevertheless, the above works have not actually been designed for processing "Big Data" sets (the data are still primarily based on limited experimental data). A Big Data conceptual architecture for cleaner manufacturing and maintenance processes of complex products was proposed [12]. Detailed design and industrial applications for specific machining processes, however, have not been reported.

Based on the literature survey and industrial survey conducted during some latest research projects by the authors, the following research gaps have been identified:

For CNC machining execution lifecycles, there are lacking systematic integration of Big Data collection, analytics and manufacturing re-scheduling optimization to address various dynamic working conditions adaptively for achieving energy efficient optimization;

Majority of the developed systems are still in laboratorial environments while industrial deployment and validation by using industrial case studies to prove the applicability of the systems to practical applications are imperative.

The goals of I^2S are: (1) to design and integrate Big Data analytics and intelligent immune mechanisms, hence to conduct condition monitoring, analysis and energy efficient optimization over manufacturing lifecycles systematically, and (2) to prove its industrial applicability through system deployment into manufacturing companies.

3 System Framework

For a machining process, it is usually managed as a series of production cycles. During each cycle, the types of components for production are certain while the quantities of each component for production could be varying. When a new production cycle starts, the types of components will be adjusted. That is, new types may be added for machining during this cycle and old types during the last cycle may be discontinued. For such a cycle, it requires optimized scheduling for multiple components to be machined in multiple machines to achieve good energy and manufacturing performance. Due to disruption and uncertainty during manufacturing execution lifecycles (e.g., dynamic changes of job priority, unexpected delay, aging or degrading of tooling and machines), it is essential to update scheduling adaptively (i.e., rescheduling) when machining conditions are changed. I^2S, which supports the above process, consists of the following functions (also shown in Fig. 1):

Vibration or acoustic sensors have been frequently used for machine condition monitoring. In this research, an electricity sensors-based Wireless Sensor Network (WSN) is integrated with CNC machines as a CPS for measuring the energy consumption of the CNC machines. The underlying consideration is that, in comparison with vibration and acoustic sensors, the developed electricity measurement sensors-based WSN is more flexible in terms of configuration and deployment. Experiments show that the energy consumption status can reflect machine and tooling conditions effectively [25]. In this research, the electricity Big Data is used to support not only energy consumption optimization but also machine condition monitoring to address dynamics of machining processes. A Big Data infrastructure has been developed for collecting, storing and processing the real-time energy data from CNC machines;

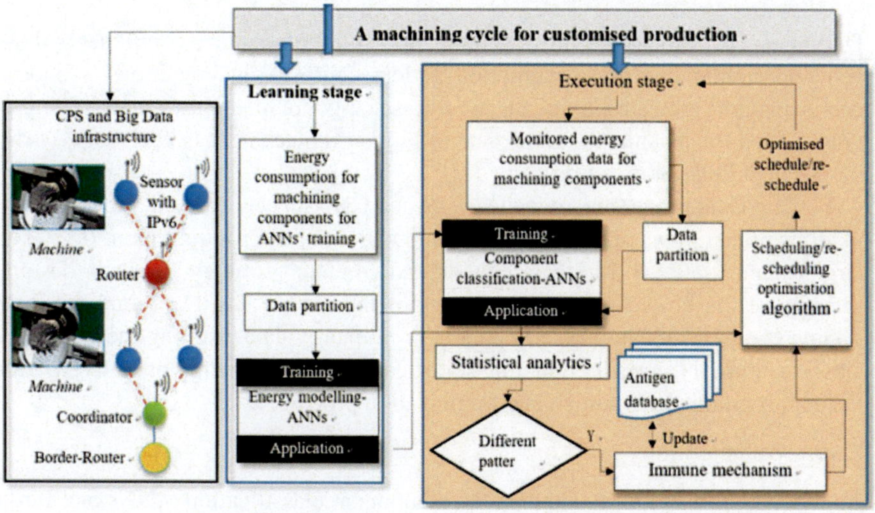

Fig. 1 I^2S for monitoring, analysing and optimizing manufacturing energy efficiency

I^2S defines an innovative architecture for manufacturing execution lifecycle. That is, a machining cycle is divided into two stages: learning at beginning, followed by execution:

(1) During the learning stage, typical components are machined to generate energy consumption patterns of machining components for ANNs' training. Two ANNs, i.e., energy modelling-ANNs and component classification-ANNs, are trained. The two ANNs are designed to model the energy consumptions for machined components and to monitor the energy patterns of the machining components respectively. A scheduling optimization, which is based on the energy modelling-ANNs, will generate an optimized scheduling plan to support the execution stage. The learning stage and the set-up of the relevant energy modelling-ANNs is reported by the authors' research group [26];

(2) During the execution stage, with the support of the component energy classification-ANNs, I^2S is used to analyze and process monitored energy data to address various dynamic conditions over machining execution lifecycles. A re-scheduling optimization will be employed to generate an optimized re-scheduling plan whenever significantly different energy patterns are identified which indicate the necessity of adjusting manufacturing plans. The details of the execution stage will be introduced in detail in this chapter.

4 Intelligent Immune Mechanism

4.1 Processing of Monitored Energy Data

During the execution stage, the machining operations of identical components should generate similar energy consumption patterns with slight deviations. Abnormal energy patterns indicate significant condition changes of machines and/or tooling. As thus, a need for machine maintenance or tooling replacement is necessary, leading to re-scheduling optimization.

For Big Data processing, deep learning and CNNs have pros and are also limited in terms of complex computing, high computational power requirement (GPU) and large training sets [7]. In I^2S, a "divide and conquer" strategy has been designed to improve the efficiency and robustness of the energy Big Data analysis: partitioning energy Big Data into component level-energy data, training the component classification-ANNs and conducting statistical analysis on energy pattern deviations to support intelligent immune and optimization processes.

Energy data partition.

As illustrated in Fig. 2, to support the component classification-ANNs and further statistical analytics for energy deviation, monitored energy data are partitioned into a

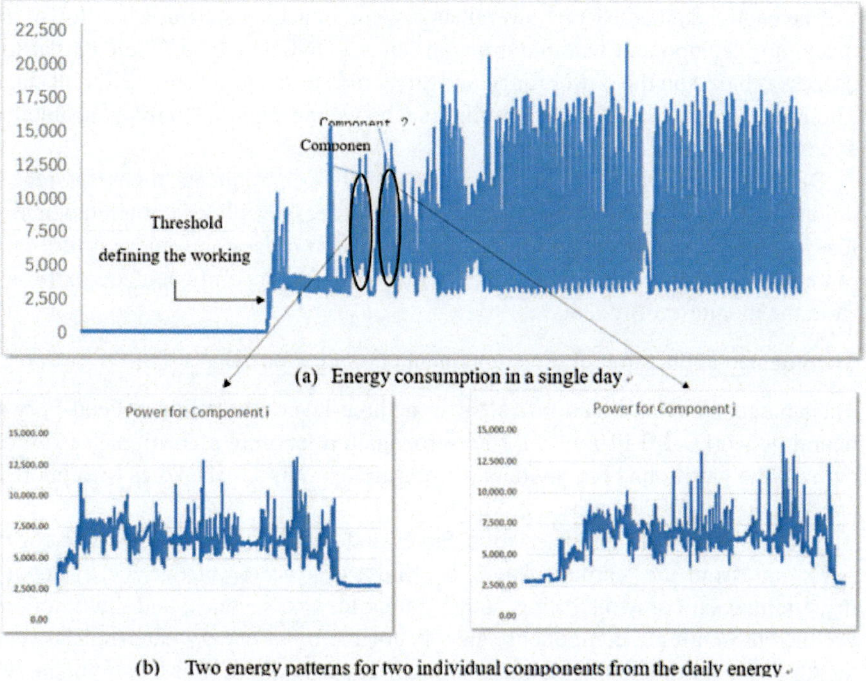

Fig. 2 Energy partition into individual energy patterns for individual components

series of energy patterns for machining individual components. The partition process is below:

Fig. 3 Design of the component energy classification-ANNs

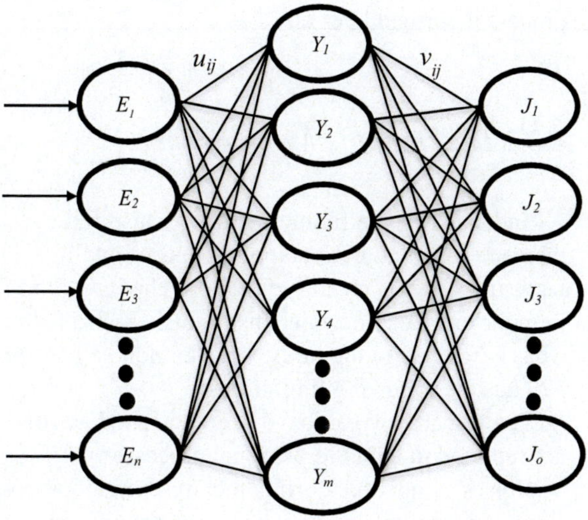

The energy data consist of several stages, e.g., machine start-up/shut-down, idle, machining, component loading/unloading into a CNC machine. The data partition process is based on the power range and thresholds to separate these different stages. The ranges and thresholds for specific machines and materials are determined via experiments;

During the process of machining two consecutive components, there is a stage for unloading the component when its machining process has been completed, and for loading a new component for machining. Based on ranges and relevant thresholds of unloading/loading, the energy data of machining each component are partitioned from the monitored Big Data.

Training and application of the component classification-ANNs.

The component classification-ANNs is a three-layer Multi-Layer Feed-Forward neural networks (MLFF) with a back-propagation learning algorithm. It is used to classify the partitioned energy data of machining components, so as to support the analysis on energy deviation conditions.

The ANNs design is illustrated in Fig. 3 and Table 2. The training data for this ANNs are from the learning stage. The input is a vector of the energy input of machining each component (e.g., three points for one second), and the output is a vector representing a component category for the input energy pattern. The vector length of the input n is the maximum length of each pattern. For a component with a smaller length, 0 will be added at the end of the pattern to standardize the vector lengths of all the patterns to be in the same length to facilitate the ANNs' processing. In terms of output, o is the total number of component types. For instance, if the output is for Component 1, the output will be [1 0 0 ··· 0]. The details of the ANNs design are explained in Sect. 5.

If a pattern for a component is judged significantly different from its standard pattern of the component, the pattern will be further processed by using the immune process described later on.

4.2 The Immune Mechanism

As indicated in the literature survey, inspired by biological immune mechanisms, there are some studies on applications of intelligent immune mechanisms for manufacturing systems recently [15, 16]. The research is based on a N/S (Non-self/Self) immune mechanism, which distinguishes what belongs to the body, i.e., the self, from what is foreign to the body, i.e., the non-self for problem identification. However, N/S has the following limitations:

The research works have been designed as conceptual frameworks and they are not reported to apply to practical processes yet.

Noises generated during the machining processes, and various dynamics of machining processes, could lead to new energy consumption patterns, which are not necessarily judged as problems. The N/S mechanism is rigid, which hinders the robustness and accuracy of the immune process.

The machining system changes over its lifetime (analogous to the human body's aging process). The N/S immune mechanism does not take into account such self-changes. A danger theory is a more flexible immune mechanism (Silva and Dasgupta, 2016). In I^{2}S, an innovative danger theory-based immune mechanism has been designed to monitor, analyze and control machining systems at the occurrence of disruption during machining processes. Key concepts and processes are below (also illustrated in Fig. 4):

Antigens—in an immune mechanism, an antigen presents the characterizing features of problems from machines, tooling and/or jobs, such as tool wear, spindle failure, machine breakdown, rush order/order modification/cancellation, control system failure, timing anomaly, etc.

Danger signals—for the temporal series of energy consumption of the machining system functioning over time (monitored Big Data), anomalies are automatically detected from the temporal analysis of energy consumption to determine whether deviations exceed tolerable variations in an energy consumption pattern. If danger signals are matched with pre-defined antigens, they are processed accordingly. For danger signals that do not match any samples in the antigen databases in its danger zone, the danger signals will be reported to engineers for problem identification. The relevant antigen databases are updated accordingly.

Danger zone—each component is associated with a danger zone, which constrains the working scope of matching between danger signals and antigen databases for the zone. Within a danger zone, the component classification-ANNs and statistical analytics are used for danger signal identification.

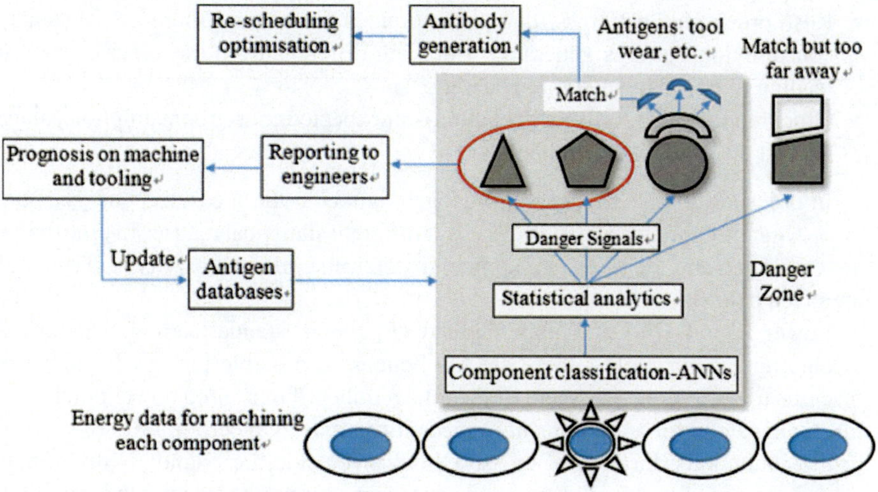

Fig. 4 The danger theory-based I^{2}S framework

Antibody and re-scheduling optimization algorithm—to handle abnormal conditions detected as antigens and carry out re-scheduling optimization if necessary.

Danger signal detection (antigen detection).

The anomaly detection process consists of two steps: (1) detecting anomalies by inputting the monitored energy data to the component classification-ANNs. If it is aligned with the standard energy patterns of corresponding components within the similarity threshold, it is judged that there is no anomaly detected and the system will do nothing. Otherwise, the newly input energy pattern is classified as anomalousness and the energy consumption pattern needs to be further analyzed in the second step; (2) comparing the pre-defined statistical metrics between the new pattern and the normal (standard) patterns to identify potential danger signals:

- Machine breakdown—no energy consumption incurred.
- Machine not working—the energy of machining a component is close to or below the standby energy for a long duration.
- Machine over-consumption and under-consumption—the power exceeds the machine-specific limits;
- Tool wear—level shift up (vertical shift up) of the energy during machining, but the energy during idle is kept unchanged;
- Cutter breakdown—energy consumption during material removal is close to that of air cutting.
- Spindle failure—energy spike unaccompanied by shift in spindle Rotation Per Minute (RPM), followed by increased energy for machining the component and then idle energy.
- Timing anomaly of control system—unexpected time shift (horizontal shift) of energy pattern between different stages.
- Rush order, new patterns—due to customized production, there are new energy consumption patterns generated which are different from the previous patterns leading to new danger signal generation.
- Order modification, order cancellation—unexpected energy consumption patterns against the original scheduled plan.

In order to identify the potential danger signals in the measured energy pattern $S_{measure}$ against a standard pattern $S_{standard}$, different statistical metrics in time domain are calculated and measures are applied to determine the anomaly type. The details are given below.

Lower bound (P_L) and upper bound (P_U) of a standard energy pattern for machining a component—these are the bounds of possible powers for a specific machine to make the component. If the total number of measured power points at the machining stage for a component exceeds its bounds by a certain level (e.g., 10%), or the continuous duration of measured power outside the bounds is more than a specific number (e.g., 5), it is treated as over-consumption or under-consumption for machining this component. The definition is given in Eq. (1).

$$\begin{cases} \frac{Numberof(S_{measure}(i,i=1:N1)>P_U)}{N1} > 10\% \rightarrow over - consumption \\ Numberof continuous points(S_{measure}(i, i = 1 : N1) > P_U) > 5 \rightarrow over - consumption \\ \frac{Numberof(S_{measure}(i,i=1:N1))<P_L)}{N1} > 10\% \rightarrow under - consumption \\ Numberof continuous point(S_{measure}(i, i = 1 : N1) < P_L) > 5 \rightarrow under - consumption \end{cases} \quad (1)$$

where N_1 is the total number of the measured power points in $S_{measure.}$

Limit of the standby duration—it is defined as the maximum continuous standby duration when the machine is treated as on normal operations. If the continuous standby duration of the measured power at the machining stage exceeds the limit, a non-working alarm will be sent out for further check (e.g., the machine door is not properly closed or a cutter is broken down). Based on the results of experiments, it is defined as 60 s in I^2S.

Cross-correction coefficient X_{coef}—cross-correlation takes one pattern ($S_{measure}$), and compares it with a shifted pattern ($S_{standard}$). The cross-covariance between the pair of patterns ($\sigma_{Smeasure,Sstandard}$) is defined below:

$$\sigma_{Smeasure,Sstandard}(T) = \frac{1}{N-1}\sum_{t=1}^{N}(S_{measure}(t-T) - \mu_{measure})((S_{measure}(t) - \mu_{standard}) \quad (2)$$

where $\mu_{measure}$ and $\mu_{standard}$ are the means of time series. N is the bigger number of the samples of the two patterns. Zero-padding the shorter pattern is to make the two patterns the same length before calculation.

The values of cross-covariance at different time shifts $\sigma(T, T = 1 : N))$ are then calculated and normalized by the variance of each pattern as the cross-correction coefficients:

$$X_{coef} = \max(X_{coef}(T, T = 1 : N)) = \frac{\sigma_{S_{measure}S_{standard}}(T, T = 1 : N)}{\sqrt{\sigma_{S_{measure}S_{measure}}(0)\sigma_{S_{standard}S_{standard}}(0)}} \quad (3)$$

The maximum cross correction coefficient X_{coef} is used to calculate the time delay.

Time delay t_{21}—Two patterns collected from different sensors or times need to be aligned first before comparison. The time delay can be found using the calculated cross-correlation below:

$$t_{21} = T when X_{coef}(T) is maximum \quad (4)$$

Mean Absolute Percentage Error (*MAPE*)—it is the average absolute percent error between the measured pattern and the standard pattern for each time period:

$$MAPE = \frac{1}{N}\sum_{i=1}^{N}\frac{|S_{measure}(i) - S_{standard}(i)|}{S_{standard}(i)} \quad (5)$$

The *MAPE* should be calculated after the two patterns are aligned based on the Time delay t_{21} and truncated to the same length *N*. Here, two bounds $MAPE_L$ and $MAPE_U$ are defined to classify the pattern based on *MAPE*. That are:

$$\begin{cases} MAPE < MAPE_L \rightarrow samepattern(standard) \\ MAPE_L < MAPE < MAPE_U \rightarrow distortionpattern(abnormal) \\ MAPE > MAPE_U \rightarrow differentpatternorshiftpattern(abnormal) \end{cases} \quad (6)$$

The two bounds $MAPE_L$ and $MAPE_U$ are determined using the F_1 score (F_1 is a measure of a test's accuracy according to the statistical theory; its range is within [0, 1]) on a cross validation dataset:

$$F_1 = 2\frac{Precison \times Recall}{Precision + Recall} \quad (7)$$

where *Precision* is the proportion of all the positive predictions that are correct and *Recall* is proportion of all the real positive observations that are correct. They are decided using the numbers of True Positive (*TP*), False Positive (*FP*) and False Negative (*FN*) when applying $MAPE_L$ and $MAPE_U$ bounds to the cross validation dataset:

$$Precsion = \frac{TP}{TP + FP} \quad (8)$$

$$Recall = \frac{TP}{TP + FN} \quad (9)$$

Various values of $MAPE_L$ and $MAPE_U$ are tested to identify the best bounds based on maximizing the *F1* score approaching 1. When $MAPE > MAPE_U$, a further calculation is needed to classify whether it is a different pattern or a vertical shift of the pattern. To this end, the measured signal is first determined using the following:

$$\begin{gathered} ME = \tfrac{1}{N}\sum_{i=1}^{N} S_{measure}(i) - S_{standard}(i) \\ S'_{measure}(i) = S_{measure}(i) - ME \end{gathered} \quad (10)$$

where *ME* is the average difference between two patterns; $S'_{measure}$ is the shifted $S_{measure}$ to $S_{standard}$.

MAPE of $S'_{measure}$ and $S_{standard}$ is calculated again to determine the anomaly type. If $MAPE < MAPE_L$, these two patterns are similar but shift vertically. If $ME > 0$, the pattern is shifted up, which may relate to a spindle failure or tool wear.

Based on the combination of the above statistical metrics, the types of danger signals can be identified by comparing the measured signal and standard signals based on the component modelling-ANNs. Detected danger signals could be from three aspects: (1) pattern deviations due to machine system problems, (2) pattern deviations due to the aging condition of the manufacturing system, (3) newly found patterns due to new types of machined components or from other aspects. Detailed steps for the immune process can be described below:

Immune process.

1. Check if the power is close to zero during scheduled machining. If yes, send "Machine breakdown" message; otherwise, go to next step;
2. Check if the power during machining is close to the air cutting power. If yes, send a "Cutter broken" message; otherwise, go to next step;
3. Use Eq. 1 to check whether there is over-consumption or under-consumption. If yes, send a "Machine over-consumption or under-consumption" message; otherwise, go to next step;
4. Count the standby duration of power during machining against the limit of standby duration. If the limit is exceeded, send a "Machine not working" message; otherwise, go to next step;
5. Employ the component classification-ANNs to check the corresponding component of the measured pattern belongs to. If the pattern belongs to the scheduled component, go to Step 7;
6. If the pattern belongs to one of the scheduled components to be machined, send an "Order modification" message and go to Step 7 for further danger signal detection; otherwise, no scheduled component is found, send a "Rush order, new pattern" message and stop;
7. Calculate the time delays t_{21} of two patterns for the same components during machining (one is the standard pattern and another one is the monitored pattern). If any time shift is more than 10 s, send a "Timing anomaly of the control system" message; otherwise, align the two patterns using the time delay and truncate the longer pattern to the same length of the shorter pattern;
8. Calculate the *MAPE* for the two patterns. If $MAPE_L < MAPE < MAPE_U$, send a "Distortion pattern" message; if $MAPE > MAPE_U$, go to next step;
9. Use Eq. 10 to determine whether the monitored pattern is a level shift from the standard pattern. If both the pattern during idle and machining is shifted up, send a "Spindle failure" message; if the pattern during machining is shifted up and that during idle is at the same level, send a "Tool wear" message;
10. Antibody stimulation will be triggered for the following conditions: (i) if the change of a cutter is needed and re-schedule when necessary (for tool wear), (ii) maintenance of spindle by finding an alternative machine and re-schedule when necessary (spindle failure), (iii) maintenance of a broken machine by identifying an alternative machine and re-schedule when necessary (machine breakdown, timing anomaly), (iv) re-schedule for a rush order;
11. Antigen databases—antigen databases will be updated if new danger signals are detected and processed as abnormal conditions (e.g., on a weekly basis or based on a factory's requirements for configuration).

4.3 Re-Scheduling Optimization

A multi-objective optimization model and algorithm for scheduling a manufacturing process have been developed by the authors' research group (Liang et al., 2018). This optimization model will be re-used for re-scheduling in this chapter. Some definitions are given below.

The energy consumption of a machine is from machining and waiting phases:

$$E_{total}(M_i) = E_{machining}(M_i) + E_{waiting}(M_i) \tag{11}$$

where $E_{total}(M_i)$ represents the energy consumed during all the phases of Machine M_i. $E_{machining}(M_i)$ and $E_{\text{waiting}}(M_i)$ represent the energy consumption of this machine during the machining and waiting phases, respectively.

The energy consumption of Machine M_i during the machining phase is computed below:

$$E_{machining}(M_i) = \sum_{j=1}^{m}(A_{ij} \times E_{machining}(M_i, J_j)) \tag{12}$$

where A_{ij} represents whether Machine M_i needs to be machined for Component J_j. $E_{machining}(M_i, J_j)$ represents the energy consumption of machining Component J_j by Machine M_i. m is the total number of components to be machined. A_{ij} can be defined as below:

$$A_{ij} = \begin{cases} 1 \quad Component\,J_j is machined by M_i \\ 0 \; Component\,J_j isnot machined by M_i \end{cases} \tag{13}$$

The total energy consumption for all the machining jobs by all the machines can be calculated below:

$$E_{total} = \sum_{i=1}^{n} E_{total}(M_i) \tag{14}$$

where E_{total} represents total energy consumption in all machines. n is the number of total machines.

To calculate the time used during the whole production time: makespan, which is the maximum production time for all components in all machines, can be computed below:

$$Makespan = \underset{j=1}{\overset{n}{Max}} \; (T_{total}(M_i)) \tag{15}$$

The balanced utilization of machines in a shop floor is defined below:

$$\mu = \frac{\sum_{i=1}^{n} T_{total}(M_i)}{n} \tag{16}$$

$$Utlisation\_level = \sqrt{\sum_{i=1}^{n} (T_{total}(M_i) - \mu)^2} \tag{17}$$

In this research, minimization of energy consumption, makespan and balanced utilization level of machines are considered. As these three objective functions have very different magnitudes, normalization of the objective functions is required. Since the maximum and minimum values of these three objective functions are unknown before optimization, a suitable normalization schema that normalizes the objective functions in the Nadir and Utopia points has been employed. The Utopia point z_i^U provides the lower bound of the ith objective function and can be obtained by minimising the ith objective function individually, i.e.,

$$z_i^U = f_i(x^i) = min\{f_i(x)\} \tag{18}$$

The upper bound is then obtained from the Nadir point z_i^N, which is defined as:

$$z_i^N = f_i(x^k) = \max_{1 \ll j \leq I}\{f_i(x^j)\} \tag{19}$$

This normalization schema may be computationally expensive when the problem dimension is very large. For this research, the time spent on this calculation is acceptable as the number of optimization parameters is not very large. Hence, the energy consumption, makespan and utilization level are to be normalized individually as:

$$\begin{cases} NE = (E_{total} - z_1^U)/(z_1^N - z_1^U) \\ NT = (Makespan - z_2^U)/(z_2^N - z_2^U) \\ NU = (Utilisation - z_3^U)/(z_3^N - z_3^U) \end{cases} \tag{20}$$

The fitness function is calculated as weighted sum of the three objectives below:

$$Fitness : min\big(w_1 \dot{N} E + w_2 \dot{N} T + w_3 \cdot NU\big), w_1 + w_2 + w_3 = 1 \tag{21}$$

where w_1, w_2 and w_3 are weights for optimization objectives.

An evolutional optimization algorithm, i.e., Fruit Fly Optimization (FFO), has been developed and improved for re-scheduling optimization. The algorithm is reported in detail in [26].

(a) Mazak machine (b) A component machined by Mazak (c) Energy measurement

Fig. 5 I^2S deployed into a Spanish company

5 Industrial Deployment and Case Studies

5.1 *Set-Up of Machining Processes and Energy Data Collection*

I^2S has been deployed in some machining companies in Sweden, U.K. and Spain (illustrated in Fig. 5 and Fig. 6). The deployment into a UK company (shown in Fig. 6) is used here for explanation. The UK company specializes on machining high-precision components for automotive, aerospace and tooling applications. A production line, consisting of three CNC machines (MX520, Mazak VC-500A 5X and Haas VF-10) and accessory equipment, has been monitored and analyzed for six months continuously to achieve energy efficient optimization.

For I^2S, a WSN is integrated with the CNC machines as CPS. The CPS is operated through radio communication—IEEE 802.15.4 (6LoWPAN (IPv6 over Low power Wireless Personal Area Networks)) and WSN producers' own communication protocols (NXP). Three-phase current and voltage are measured by utilizing current and voltage sensors. Measured data is transmitted through the 2.4 GHz Wi-Fi to a coordinator connected with an Internet-router. Energy data is stored in Hadoop for Big Data processing. For each CNC machine, the data rate of energy data for each machine is 3 readings per second. Some of the production line and monitored data are shown in Fig. 6.

5.2 *Big Data Partition and ANNs Training*

The machining process was carried out six days per week. During production, the stages include machine start-up/shut-down, idle, machining, etc. On Sunday, the machine is set either idle or shut-down. Before the deployment of I^2S and rescheduling optimization, the energy consumption for a week is shown in Fig. 7. For the machining process between consecutive components, there are an unloading stage of a component that has been already machined, and a loading stage of raw

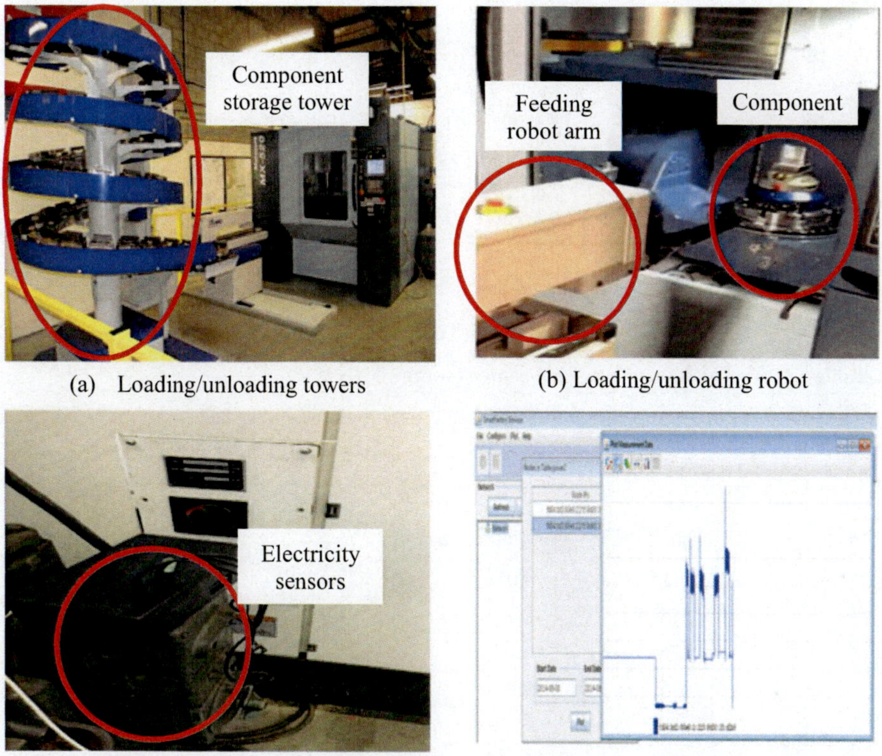

(a) Loading/unloading towers (b) Loading/unloading robot

(c) Power measurement (d) Energy measurement result

Fig. 6 Energy measurement on CNC machines in a UK company

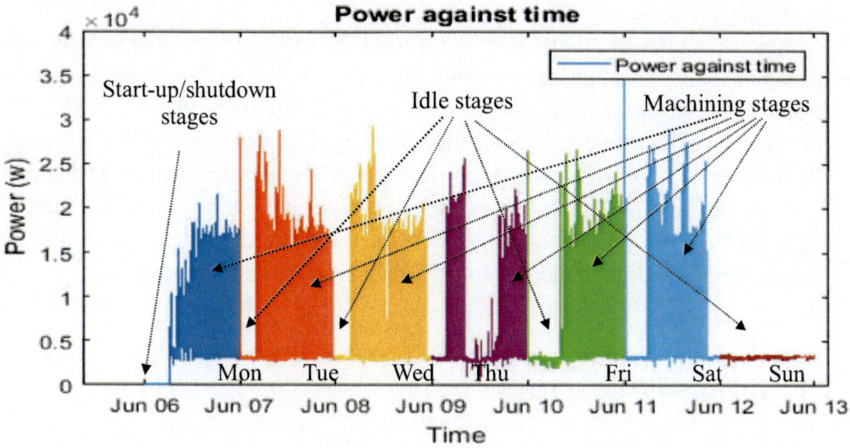

Fig. 7 Energy consumption monitored for a week (before the re-scheduling)

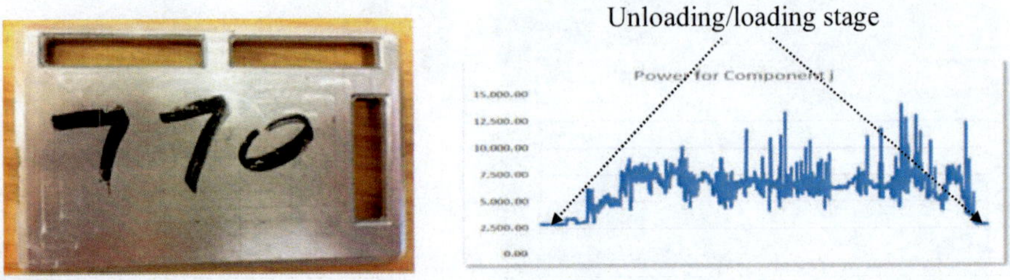

Fig. 8 Energy consumption monitored during the unloading/loading stages

Table 3 Different middle layers and neurons for the ANNs

Number of middle layers	Number of neurons in middle layers	Average training time (s)	Average iterations (times)	Average accuracy (%)
1	509	1.973	26.6	98.97
2	509	3.244	30.2	98.69
1	1018	4.018	27.6	97.69
2	1018	8.475	31	96.92
1	2036	6.711	26	98.46
2	2036	19.964	23.6	96.92
1	2037	7.703	29.2	98.97
2	2037	24.169	30.6	98.20

materials for machining another component. The unloading/loading energy profiles are shown in Fig. 8.

In I^2S, 13 types of components with total 78 energy patterns have been utilized to train the component classification-ANNs. Different structures have been used for comparison (shown in Table 3). The structure of a middle layer with 509 neurons in this layer has been finally selected due to its higher efficiency and more accuracy. The number of neurons for the input layer is 1018, and the number of neurons for the output layer is the number of component types, which is 13 in this study.

Figure 9 shows some results by the trained ANNs (last one is the ANNs model trained using the aligned dataset (only 3 data samples are potted here).

5.3 *Immune Processes*

Ranges of powers for different stages were measured for individual machines (e.g., for MX520, the power for idle/loading/unloading is about 3.5–3.7 kw, for machining it is about 8.0–15 kw, for air cutting it is about 6.0–8.0 kw). There are some danger signals (potential issues) to be processed by the immune mechanism of I^2S. Before

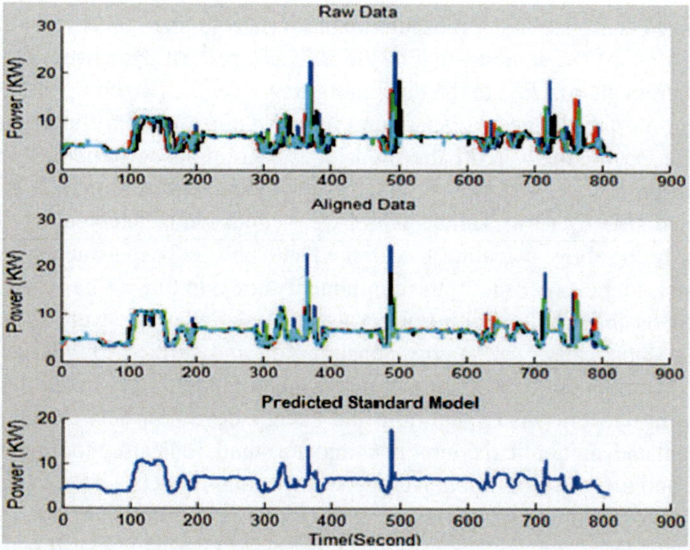

Fig. 9 Training of the ANNs for identical components

various danger signals are identified, the lower and upper bounds for detection of new or distortion patterns ($MAPE_L$ and $MAPE_U$) are determined by calculating the F_1 score and trying various values of $MAPE_L$ and $MAPE_U$ on a group of a cross validation dataset. The best ones are selected based on the F_1 score. In this experiment, $MAPE_L$ and $MAPE_U$ are determined as 0.09 (9%) and 0.20 (20%) respectively.

Case 1: Tool wear.

Tool wear detection is illustrated in Fig. 10. The machining processes for two identical components are compared. The energy consumption for the component under an ideal condition is defined as a standard pattern (in blue) and the pattern to be inspected is defined as a measured pattern (in red). In the standard pattern (in blue), its idle stage lasts from 0 to 50 s with a power level of 2.94KW, and the machining stage

Fig. 10 Energy pattern deviation due to tool wear

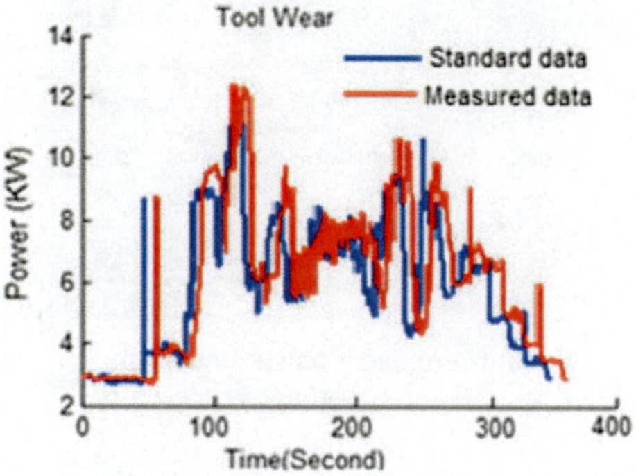

runs from 51 to 355 s. In the measured pattern (in red), the power level at idle stage (0–56 s) remains the same as that of the standard pattern. However, the machining energy increment (*MAPE*) in the machining stages for the two components is 21.8%, exceeding the pre-defined upper bound $MAPE_U$ (20%). To further determine the deviation type, the measured pattern during machining is de-trended using Eq. (10) to obtain the $S'_{measure}$. *MAPE* of $S'_{measure}$ is calculated as + 6%. It is below the pre-defined $MAPE_L$ (9%). Hence, a tool wear condition is detected.

Actually, thc above calculation is aligned with physical experiments conducted by the authors. In the experiment, the component shown in Fig. 11 has been repeatedly machined on the CNC machine for a number of times to ensure the conclusion is correct statistically. The energy consumption and surface roughness for all the machined component have been measured and compared. It is noted that when the cutter is in a severe wear condition, the energy deviation between the measured component and an identical component under a standard healthy tooling condition is 21.67%, and the corresponding average surface roughness is 1.15 μm and 3.22 μm respectively (shown in Fig. 12). For healthy tooling conditions, the energy deviation is far below 18% and surface roughness is below 1.15 μm. The threshold (18%) obtained using the experiments is very close to $MAPE_U$ (20%) that is optimized based on the F_1 score. Therefore, the approach is validated through the experiments.

Case 2: Timing anomaly.

For timing anomaly, the pattern difference is shown in Fig. 13. An unexpected time delay of 35 s at No. 480 s in the standard pattern and No. 505 s in the measured pattern is detected. Such long-time shift is more than timing anomaly threshold so that it could indicate a possible timing anomaly.

Case 3: No working.

No working will generate long standby as shown in Fig. 14. The measured power is around the standby power (3.6 KW for the machine used in this case) for a very long period (as defined previously, if the standby time exceeds 60 s, a warning message

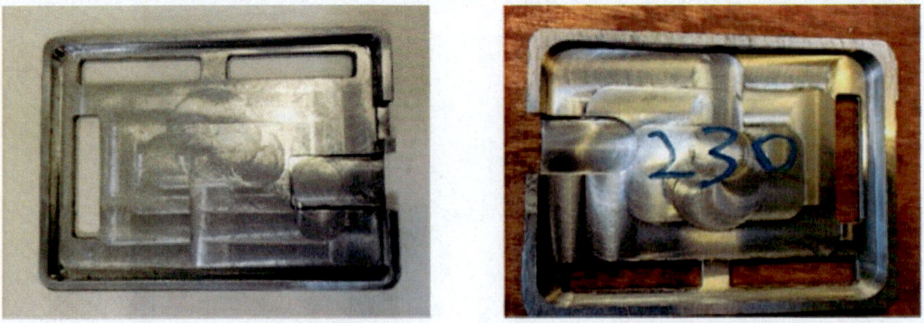

(a) Component under a standard condition

(b) Component under an abnormal condition

Fig. 11 A component is machined under standard and abnormal conditions

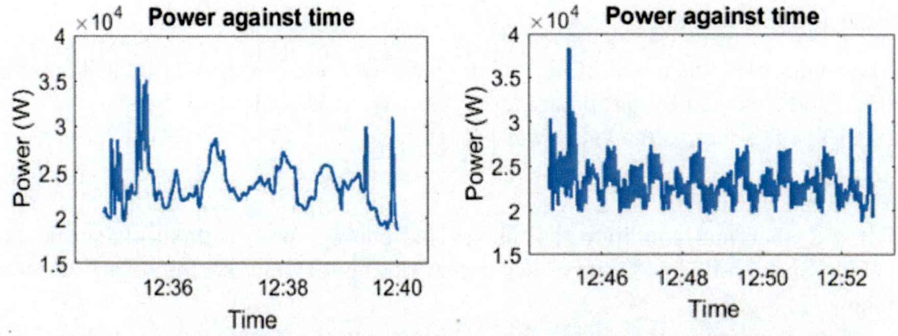

(a) Power profile under a standard condition

(b) Power profile under an abnormal condition

Fig. 12 Energy profiles for components under standard/abnormal conditions (energy consumptions are 2.03 Kwh and 2.47 Kwh respectively—deviation is 21.67%)

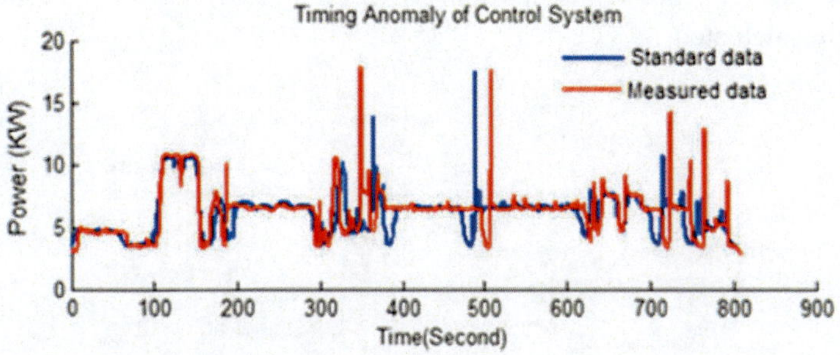

Fig. 13 Energy pattern deviation due to timing anomaly

Fig. 14 Energy pattern deviation due to a non-working condition

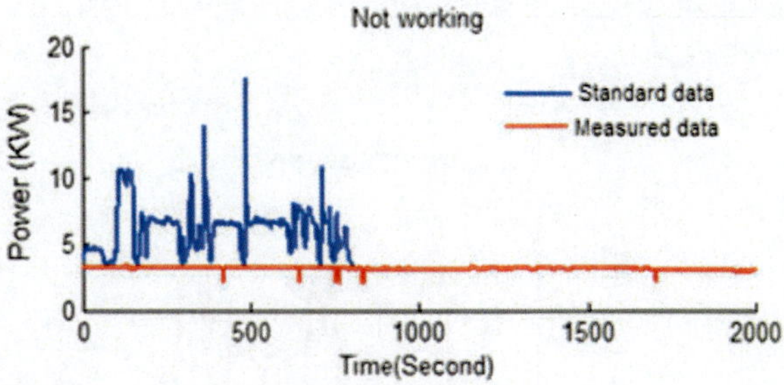

would be sent out). It indicates the machine is not working. It is found out that the machine door was not closed properly so that the machine was not running.

Case 4: Cutter broken.

The situation is illustrated in Fig. 15. It starts from the No. 214 s (the spike in red line), and followed by the power level (3.9 kW) that is close to the standby level (3.6 kW). It is judged the cutter was broken from No. 214 s.

Case 5: Spindle failure.

For this abnormal condition, an unexpected energy spike without change in the spindle speed followed by increased energy during idle and machining are detected (shown in Fig. 16).

In the standard pattern (in blue line), the idle stage lasts from No. 381 to No. 470 s at a power level of 3.6KW. The machining stage lasts from No. 471 to No. 760 s. In the measured data pattern (in red line), an unexpected energy spike is detected in No. 351 s and the following energy during idle and machining shifts up at levels of 21.5% and 20.3% respectively. They exceed the pre-defined threshold 20%. As thus, the measured data is de-trended using Eq. (10) and *MAPE* of the de-trended dataset is +4.3% (which is below the pre-defined $MAPE_L$ (9%)). As thus, a spindle failure condition is detected.

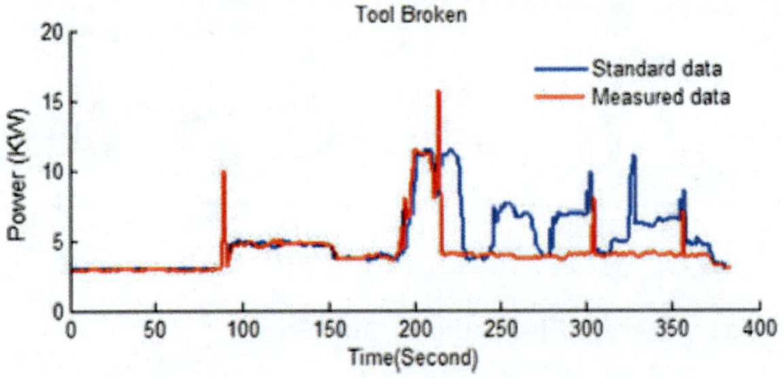

Fig. 15 Energy pattern deviation due to a broken cutter

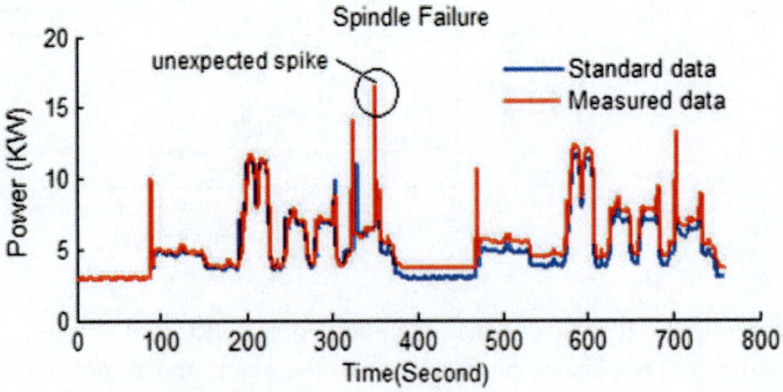

Fig. 16 Energy pattern deviation due to a spindle failure condition

Case 6: New patterns.

It is judged that the signal in Fig. 17 is a new component for machining. Any abnormal conditions are detected in these patterns. As thus they are recorded as normal patterns and will not be handled as a danger signal. The newly generated patterns for machining new components are updated into the training set of the component classification-ANNs. Danger signals shown in Figs. 18 and 19 are new signals and will be reported to maintenance engineers. If judged as abnormal, they will be processed and recorded into the antigen databases.

System condition evolving.

The thresholds of machines are adjusted based on experiments to be adaptive to the machine aging conditions. The relevant thresholds should be set based on longer term experiments (the research work is ongoing).

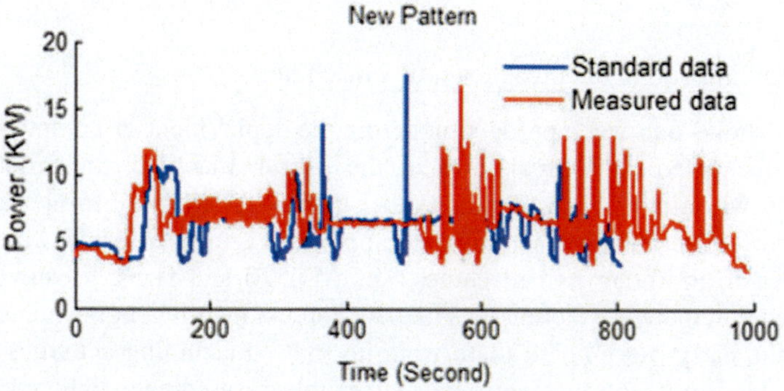

Fig. 17 A new energy consumption pattern for a newly machined component

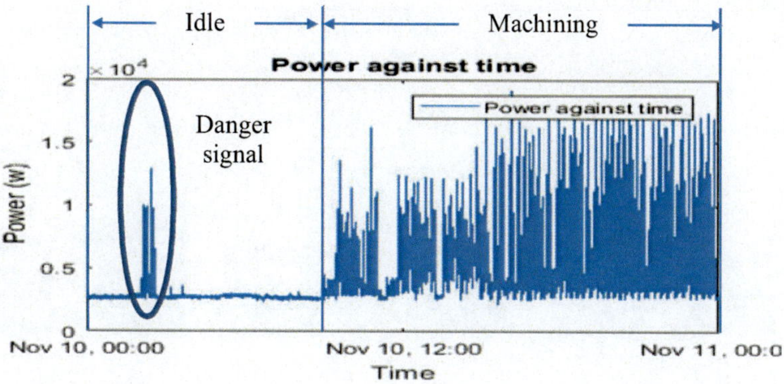

Fig. 18 A danger signal

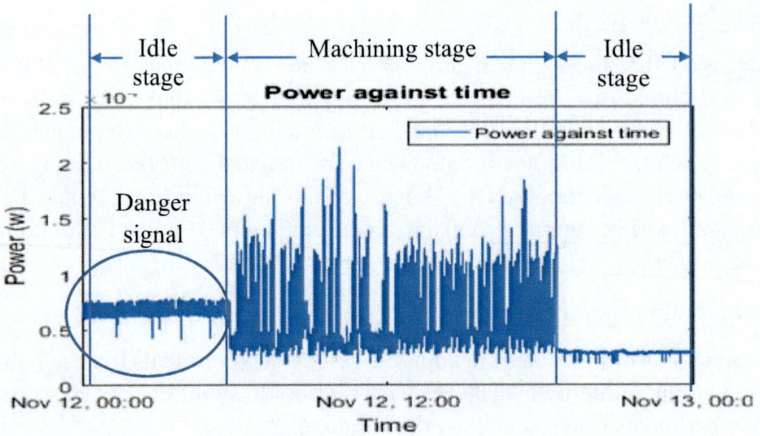

Fig. 19 A danger signal

Re-scheduling due to the breakdown of a machine.

Figure 7 shows one-week production before the deployment of I^2S into the factory. Due to the unexpected breakdown of the Mazak machine, the productivity on Thursday was almost 50% less than other days. Via I^2S, the Mazak machine was identified for maintenance so that the components arranged for the Mazak need to be shifted to the other two machines, i.e., MX520 and Haas. Meanwhile, it was identified that there are more issues like door stuck, insufficient raw material supply during the early morning so there were no many machining activities during the periods. Measurements were introduced to enable the automatic lines to keep continuous execution during the early mornings as well. Re-scheduling optimization made improvements for energy saving and productivity (the three-month improvement in the factory is shown in Fig. 20).

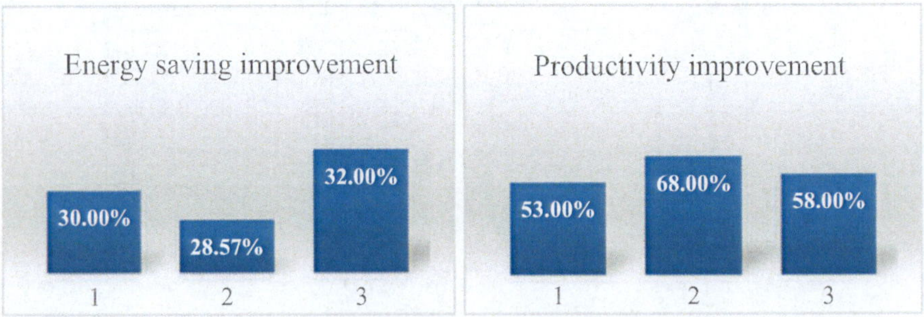

Fig. 20 Improvements for three months by using I^2S into the production line

6 Conclusions

In this chapter, an innovative I^2S has been developed for energy efficient manufacturing monitoring, analysis and optimization. The system is enabled by effective Big Data analytics and intelligent immune mechanisms to adapt to condition changes of machining processes. A re-scheduling algorithm is triggered when necessary thereby achieving multi-objective optimization of energy consumption and manufacturing performance. The innovations and characteristics of I^2S include:

By integrating with CPS, ANNs and re-scheduling optimization, an innovative immune mechanism has been designed to effectively process energy Big Data to achieve energy efficient optimization for manufacturing lifecycles;

I^2S has been deployed into some European factories for trials for months. Real-world case studies have been used for system validation. For the companies, around 30% energy saving and over 50% productivity improvement have been achieved. The effective applicability of I^2S to industrial environments has been proved. By using I^2S, sustainability improvements on the environmental, economic and social aspects have been achieved (environmental and economic—energy efficient manufacturing, less unexpected breaks to improve energy efficiency and productivity, social—intelligent monitoring and processing to avoid tedious human intervention and errors as a more user—friendly working environment).

There is ongoing research to improve I^2S further. Near future research issues include:

- In practices, the error of measured data could be a very significant issue. Measures to eliminate the error effect are under investigation. Longer-term experiments will be carried out to establish more reliable thresholds of machine aging. Meanwhile, optimization approaches are under research to define the lower and upper energy bounds of machined components.
- Investigations on integrating the immune mechanism and edge computing (a new IT infrastructure) are carried out to improve the efficiency and effectiveness of I^2S for industrial applications.

References

1. Stark R, Seliger G, Bonvoisin J (2017) Sustainable manufacturing. Springer
2. Engert S, Rauter R, Baumgartner R (2016) Exploring the integration of corporate sustainability into strategic management: a literature review. J Clean Product 112(4):2833–2850
3. ElMaraghy H, Nassehi A (2016) Computer-aided process planning. Springer, CIRP Encyclopedia of Production Engineering
4. Wang S, Lu X, Li XX, Li WD (2015) A systematic approach of process planning and scheduling optimization for sustainable machining. J Clean Product 87:914–929
5. Monostori L, Kádár B, Bauernhansl T, Kondoh S, Kumara G, Sauer O, Sihn W, Ueda K (2016) Cyber physical systems for manufacturing. CIPR Annal 65(2):621–641

6. Yann L, Bengio Y, Hinton G (2015) Deep learning XE "Deep learning". Nature 521(7553):436–444
7. Najafabadi MM, Villanustre F, Khoshgoftaar TM, Seliya N, Wald R, Muharemagic E (2015) Deep learning applications and challenges in big data analytics. J Big Data 2(1) (on-line open access)
8. Wang LH, Shen WM (2007) Process Planning and Scheduling for Distributed Manufacturing, Springer
9. Li W.D., McMahon C.A., 2007. A simulated annealing – based optimization approach for integrated process planning and scheduling. International Journal of Computer Integrated Manufacturing. 20(1): 80–95.
10. Tang D, Dai M, Salido M, Giret A (2016) Energy-efficient dynamic scheduling for a flexible flow shop using an improved particle swarm optimization. Comput Ind 81:82–95
11. Liu Y, Dong H, Lohse N, Petrovic S (2016) A multi-objective genetic algorithm for optimisation of energy consumption and shop floor production performance. Int J Prod Econ 179:259–272
12. Yan J, Li L, Zhao F, Zhang F, Zhao Q (2016) A multi-level optimization approach for energy-efficient flexible flow shop scheduling. J Clean Product 137:1543–1552
13. Xu W, Shao L, Yao B, Zhou Z, Pham D (2016) Perception data-driven optimization of manufacturing equipment service scheduling in sustainable manufacturing. J Manufact Syst 41:86–101
14. Salido M, Escamilla J, Barber F, Giret A (2017) Rescheduling in job-shop problems for sustainable manufacturing systems. J Clean Product 162:121–132
15. Cai Q, Tang DB, Zhu H (2016) Research on the immune monitoring model of organic manufacturing system. Proceedings of 9th International Conference on Digital Enterprise Technology. pp 533–538
16. Darmoul S, Pierreval H, Hajri-Gabouj S (2013) Handling disruptions in manufacturing systems: An immune perspective. Eng Appl Artif Intell 26(1):110–121
17. Metan G, Sabuncuoglu I, Pierreval H (2010) Real time selection of scheduling rules and knowledge extraction via dynamically controlled data mining. Int J Prod Res 48(23):6909–6938
18. Bennane A, Yacout S (2012) LAD-CBM; new data processing tool for diagnosis and prognosis in condition-based maintenance. J Int Manufact 23(2):265–275
19. Kusiak A, Verma A (2012) Analyzing bearing faults in wind turbines: A datamining approach. Renew Energy 48:110–116
20. Purarjomandlangrudi A, Ghapanchi AH, Esmalifalak M (2014) A data mining approach for fault diagnosis: An application of anomaly detection algorithm. Measurement 55(1):343–352
21. Fang K, Uhan N, Zhao F, Sutherland J (2011) A new approach to scheduling in manufacturing for power consumption and carbon footprint reduction. J Manufact Syst 30(4):234–240
22. He Y, Liu F, Wu T, Zhong F, Peng B (2011) Analysis and estimation of energy consumption for numerical control machining. Proc Inst Mech Eng, Part B: J Eng Manufact 226(2):255–266
23. Winter M, Li W, Kara S, Herrmann C (2014) Determining optimal process parameters to increase the eco-efficiency of grinding processes. J Clean Product 66:644–654
24. Yan J, Li L (2013) Multi-objective optimization of milling parameters – the trade-offs between energy, production rate and cutting quality. J Clean Product 52(1):462–471
25. Zhou LR, Li JF, Li FY, Meng Q, Li J, Xu XS (2016) Energy consumption model and energy efficiency of machine tools: a comprehensive literature review. J Clean Product 112(5):3721–3734
26. Liang YC, Lu X, Li WD, Wang S (2018) Cyber Physical System and Big Data enabled energy efficient machining optimization. J Clean Product 2018(187):46–62

Adaptive Diagnostics on Machining Processes Enabled by Transfer Learning

Y. C. Liang, W. D. Li, S. Wang, and X. Lu

1 Introduction

Throughout machining process lifecycles, machining equipment and cutting tools are prone to aging and degrading, resulting in dimensional and geometric deviations on machined work-pieces. Meanwhile, abnormal conditions could generate severe vibration or high temperature, leading to gradual or abrupt production failures. Due to the issues, higher scrap rate, greater material waste and lower productivity will be generated. Thus, effective fault diagnostics and predictive maintenance for machining process lifecycles are imperative to minimize defects, and thereby attain the enhancement of product quality, improvement of productivity and reduction of energy and material waste. In recent years, based on the significant advancement in sensor network, artificial intelligence and Big Data analytics, data driven approaches for fault diagnostics have been actively developed.

Deep learning algorithms usually work well to support machining process lifecycles. A diagnostics approach well-trained for one machining condition could not be applicable to a new condition effectively. For instance, when a machine during a machining process needs maintenance, an alternative machine is considered for replacement in order to carry on further machining processes. In machining factories, which are mainly composed of Small and Medium-sized Enterprises (SMEs), a replaced machine is usually in a different type and the machining set-ups is changed. Accordingly, a deep learning enabled diagnostics approach, which has been trained for the previous machining condition, needs to be re-trained on this new arrangement from scratch. Nevertheless, it is challenging and time-consuming to collecting a large

Y. C. Liang · W. D. Li (✉) · S. Wang · X. Lu
Faculty of Engineering, Environment and Computing, Coventry University, Coventry, UK
e-mail: weidong.li@coventry.ac.uk

W. D. Li
School of Logistics Engineering, Wuhan University of Technology, Wuhan, China

W. Li et al. (eds.), *Data Driven Smart Manufacturing Technologies and Applications*, Springer Series in Advanced Manufacturing,
https://doi.org/10.1007/978-3-030-66849-5_4

volume of labeled data and re-train the approach to ensure the high accuracy of the diagnostics function on the new condition.

Transfer learning is a promising strategy to address the challenge. The strategy is to retain and reuse the invariant knowledge and features learn in one domain (called the source domain) to improve the performance of another domain (called the target domain) [1]. Knowledge is transferred on the basis of resolving the dissimilarities of feature distributions across different domains. Transfer learning has been increasingly applied in various research areas, such as seizure classification [2], tool tip dynamics prediction [3], and image classification [4]. It is sensible to consider incorporating a transfer learning mechanism into a deep learning algorithm for designing an adaptive diagnostics approach for machining process lifecycles. That is, training knowledge and features under a machining condition (considered to be a source domain) will be adaptively transferred to the new machining condition (considered to be a target domain), and thereby significant efforts for algorithm re-training from scratch can be minimized. However, there is still no related research on this aspect reported so far.

In this chapter, transfer learning is integrated with a LSTM-CNN as an innovative adaptive diagnostic system to support machining process lifecycles. During the process, the LSTM-CNN is constructed and trained for online diagnostics on machining conditions. Furthermore, the LSTM-CNN is systematically enhanced by incorporating a transfer learning strategy as a deep transfer learning approach, to be applicable to other machining process adaptively. The characteristics and innovations of the approach include below.

(1) According to the best of our knowledge, it is a new attempt to design a deep transfer learning approach to perform diagnostics on machining process lifecycles in an adaptive means. New optimization mechanisms are devised to minimize the differences of the datasets as well as the feature distribution discrepancies on different machining conditions. The weights of the LSTM-CNN trained for a machining condition are transferred to train the LSTM-CNN for new machining conditions to ensure better training accuracy and efficiency. Meanwhile, effective mechanisms are integrated to avoid overfitting during the training processes. Based on that, adaptive diagnostics on machining lifecycles is achieved.

(2) To showcase the superior performance of the approach, comparative evaluations on various parameters, settings and algorithm design are conducted. Case studies indicate that the approach presented in this chapter achieved 96% in training accuracy, which is significantly higher than other approaches.

2 Literature Survey

2.1 Related Work on Fault Diagnostics

Diagnostics on the working conditions of equipment and manufacturing processes has been an active research topic over decades. In recent years, owing to the rapid progress of artificial intelligence, a new trend is to leverage relevant algorithms to design diagnostics functions. The related works were summarized by several survey papers [5–7]. For the approaches, multivariable signals from manufacturing systems, such as electrical current, vibration or acoustic emission, were continuously collected. Intelligent analytics were carried out in the time domain, the frequency domain, or the time–frequency domain. For instance, a stochastic-resonance based adaptive filter was designed for online fault diagnosis on motor bearings [5]. A polynomial-time algorithm for the verification of k-reliable prognosability for decentralized fault diagnosis/prognosis was presented [6]. The principal work-piece analysis (PCA) and k-Nearest Neighbor (k-NN) based method were used to detect faults for bearing systems [8], manufacturing processes [9], or the grading level of manufacturing systems [10]. Han et al. designed the Least Squares Support Vector Machine (LS-SVM) model to identify faults on centrifugal chillers. Feature extraction was implemented to acquire eight fault features from 64 raw indicators, and the overall accuracy can achieve more than 99% [11]. Suo et al. implemented Fuzzy Bayes Risk (FBR) to achieve fault diagnostics on satellite power systems [12]. In the research, a feature selection method was also proposed to support the diagnostics accuracy, and the achieved accuracy of the method is better than the other state-of-the-art systems. According to the research by Luo et al. [13], frequency features were extracted from vibration signals using a state-space model. A Cosine distance was designed to compare healthy frequency features and monitored features to detect early faults for a machining center. In recent years, deep learning algorithms, including CNN, RNN, Stacked Auto Encoders (SAE), etc., were effectively applied for fault detection on manufacturing equipment or processes. Zhong et al. integrated CNN and SVM to diagnose faults for gas turbines [14]. In the research, the overall diagnostics accuracy can achieve 93.44%. Li et al. developed a Sparsity and Neighborhood Preserving Deep Extreme Learning Machine (SNP-DELM) to detect rotating machinery faults based on motor current signals [15].

As aforementioned, the limitation of using deep learning algorithms is in their re-training processes, which require re-collecting and re-annotating a large volume of data. It is challenging to acquiring such large annotated datasets for new set-ups [16]. This unfavorable characteristic makes the algorithms less suitable in machining process lifecycles, in which production reconfigurations on machines/cutting tools could be frequent and re-training processes of deep learning based diagnostics are extensively time-consuming and costly.

2.2 Related Work on Transfer Learning Based Applications

Transfer learning has presented its potential in various applications. With transfer learning, knowledge acquired from one domain might be retained and reused in a new domain. The strategy can be designed for fault diagnostics (e.g., fault feature identification and fault classification) during machining process lifecycles to alleviate re-training. In general, the methodologies of transfer learning can be classified into the three general settings, i.e., inductive, transductive and unsupervised [1]. Under these three settings, according to what and how the knowledge is transferred, the developed approaches can be further classified as four categories: (1) Instance based transfer learning—the dataset in the source domain is re-weighted in an attempt to correct marginal distribution differences. These re-weighted instances are then used in the target domain for training [17, 18]; (2) Feature based transfer learning - the knowledge used for transfer across domains is encoded into the learned feature representation to improve the performance of the target task [19]; (3) Parameter based transfer learning—the setting parameters of an intelligent algorithm in the source domain are re-used in the target domain [20]; (4) Relational knowledge-based transfer learning—the relationship between the data from the source domain and the target domain is established, which is the base for knowledge transfer between domains [21]. For transfer learning, many approaches work well under a prerequisite condition: the cross-domain datasets are drawn from the same feature space under the same distribution. When the feature distribution changes, most deep learning based approaches need to be re-trained from scratch.

Based on transfer learning, various applications were developed. Chen et al. used instance based transfer learning to achieve tool tip dynamics prediction for new tools without conducting a huge amount of impact tests [3]. The accuracy of prediction achieved over 98%. Zhang et al. utilized instance based transfer learning to achieve image classification across different domains [22]. The accuracy of the overall cross-domain object recognition was 80.7%, and the approach was robust to data noise. On the other hand, the training accuracy can only reach 53.0 $\pm$ 0.8% by using other learning methods. Kim and MacKinnon used parameter based transfer learning to detect fracture in computed tomography images [23]. More than 95.4% accuracy was achieved, in comparison with 94% accuracy if without the developed transfer learning. Furthermore, some research works incorporated transfer learning into deep learning algorithms to set up deep transfer learning algorithms. Lu et al. developed a deep transfer learning algorithm based on deep neural network and feature based transfer learning for domain adaptation [24]. In the research, the distribution difference of the features between the datasets from the source domain and target domain was evaluated based on the Maximum Mean Discrepancy (MMD) in a Reproducing Kernel Hilbert Space (RKHS). Weight regularization was carried out by an optimization algorithm to minimize the difference of the MMD across domains in implementing knowledge transfer. Xiao et al. designed another deep transfer learning algorithm for motor fault diagnostics [25]. In the algorithm, a feature based transfer learning method was developed to facilitate knowledge l. MMD was incorporated into the training process to impose constraints on the parameters of deep learning to minimize distribution mismatch of features between two domains. Wen et al. proposed a novel deep transfer learning algorithm for fault diagnosis [26]. A cost function of weighted fault classification loss and MMD was used to fine tune the weights of the model. Sun et al. designed another hybrid parameter/feature deep transfer learning algorithm for predicting Remaining Useful Life of cutters (RUL) [27]. The weights of a trained SAE for RUL were transferred to a new SAE for predicting the RUL of new cutters. For the new SAE, the difference of features in the middle layers for the dataset in this application and the data for the previous SAE was then compared. Gradients for weights and bias in the new SAE can be solved by minimizing the divergence so that the features can be transferred. The limit of this approach is that the discrepancy of the data distribution for the two SAEs was not considered. A CNN based transfer learning approach was proposed for fault diagnosis [43], but the accuracy of training could be improved.

According to the characteristics of transfer learning, it could be an effective strategy to transfer knowledge between varying conditions throughout the lifecycle of a machining process, in which production configurations and working conditions could be kept changing. Nevertheless, no relevant research in this area are reported so far. Thus, it is timely to conduct the research of leveraging the transfer learning strategy in facilitating adaptive diagnostics on machining processes, thereby achieving improved performance in training efficiency and accuracy.

3 System Framework

3.1 Manufacturing System and Machining Process Lifecycle

A manufacturing system and machining process lifecycle are modelled here to better explain the developed adaptive diagnostics approach.

A manufacturing system can be represented below:

$$S = \prod\nolimits_{i=1}^{I} M_i\left(\prod\nolimits_{j=1}^{J} T_j\right) \tag{1}$$

where S represents a manufacturing system in a factory, which consists of a series of machines in the system (M_i – No. i machine, ($i = 1 \ldots I$)); each machine is composed of a set of associated cutting tools (T_j – No. j cutting tool for M_i, ($j = 1 \ldots J$)).

For a specific job (a group of work-pieces) to be machined, the machining process can be represented below:

$$MP_i = \left(M_i\left(\prod\nolimits_{j=1}^{J} T_j\right), \prod\nolimits_{k=1}^{K} C_k\right) \tag{2}$$

where MP_i represents the machining process performed by a specific machine Mi (No. i machine in the above manufacturing system S, $i \in I$), which is used to machine a set of work-pieces (C_k – No. k work-piece, ($k = 1,\ldots,K$)).

During a machining process lifecycle, the machining process will be changed when the executive machine is diagnosed malfunctional and a replaceable machine will be used. That is:

$$MP_i \rightarrow MP_{i\_r} \tag{3}$$

where the machining process MP_i, which is conducted using Machine M_i and its cutting tools, is changed to MP_{i_r}, which is conducted using a replaced Machine M_{i_r} and its cutting tools.

3.2 Flow of Adaptive Diagnostics

In this research, for an entire machining process lifecycle, power signals are collected to support adaptive diagnostics. According to the research of Liu et al. [28] and Sealy et al. [29], power signals can indicate the working conditions of machines and cutting tooling. Experiments showed that power sensors were easy to deploy and the data sampling rate was much lower than that of vibration or acoustic sensors [30]. The functions of adaptive diagnostics are below (illustrated in Fig. 1):

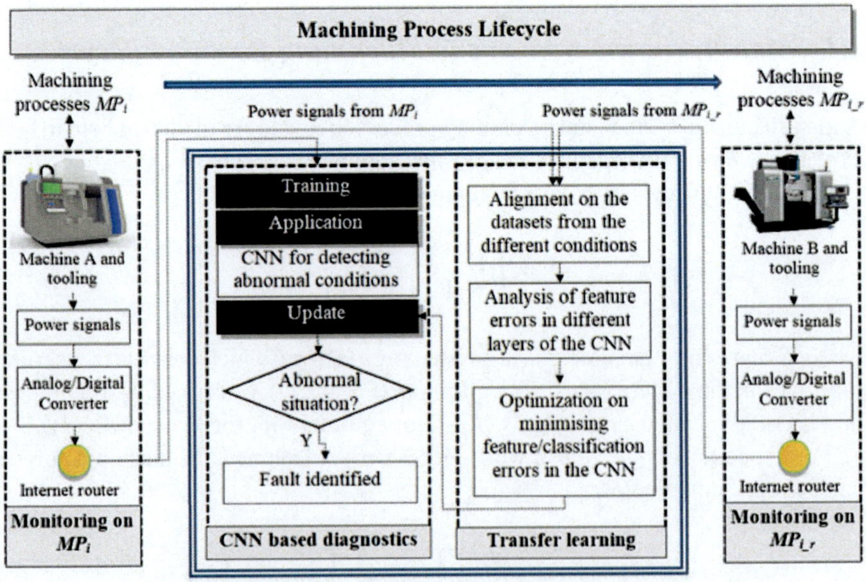

Fig. 1 Significant paradigm shifts of the global manufacturing industry

(1) Process monitoring for the current working condition (e.g., MP_i): The machine is deployed with sensor devices, circuits and routers. During MP_i, the power signals of the machine are continuously collected via power sensors and transmitted for further processing in the following step (2);
(2) LSTM-CNN based diagnostics: A LSTM-CNN is designed for detecting the abnormal working condition of the machine and associated tooling (e.g., severe tool wear, tool breakage, spindle failure of the machine). The LSTM-CNN is trained using labeled data collected for MP_i;
(3) Process monitoring for a new working condition (e.g., MP_{i_r}): During a machining process lifecycle, the machining process could be changed from MP_i to MP_{j_r} when MP_i is diagnosed malfunctional. During MP_{j_r}, the power signals are continuously collected via power sensors and transmitted for further processing in the following Step (4);
(4) Transfer learning for adaptive diagnostics: the above two machining processes are considered the source domain (MP_i) and the target domain (MP_{i_r}), respectively. That is, the diagnostics knowledge embedded in the trained CNN for MP_i (the source domain) will be transferred to MP_{j_r} (the target domain) in implementing adaptive diagnostics.

For the above functions, the implementation steps are explained further as follows (also illustrated in Fig. 2):

Step 1: LSTM-CNN designfor diagnostics on machining process.

The current machine process is MP_i. Labeled data for MP_i are collected to train the CNN for process diagnostics;

Step 2: Adaptive diagnostics on machining process lifecycle.

During the machining process lifecycle, Machine M_i is diagnosed faulty, and replaced by Machine M_{i_r}. The process of MP_i is therefore changed to the machining process MP_{i_r}. To accelerate the training efficiency for MP_{i_r}, the knowledge for diagnosing MP_i will be transferred to be applied to MP_{i_r} using the following sub-steps:

- *Step 2.1:* Optimization for dataset alignment: MP_i is considered the source, and MP_{i_r} is the target. Alignment on the datasets across the domains is performed to facilitate an effective transfer learning process. An optimization algorithm is designed to enhance the alignment;
- *Step 2.2:* Optimization for minimizing feature distribution discrepancy and fault classification error: (1) forward computation of the CNN is carried out by using both datasets of MP_i and MP_{i_r}. Based on the datasets in the two domains, computation of the feature distribution discrepancy in each layer of the LSTM-CNN is conducted; (2) back propagation computation in the LSTM-CNN is performed to minimize the feature distribution discrepancy computed in the above forward computation and fault classification error. An improved optimization process is designed for this minimization process. The updated LSTM-CNN is deployed for the new condition. The procedure goes back to the starting point of Step 2

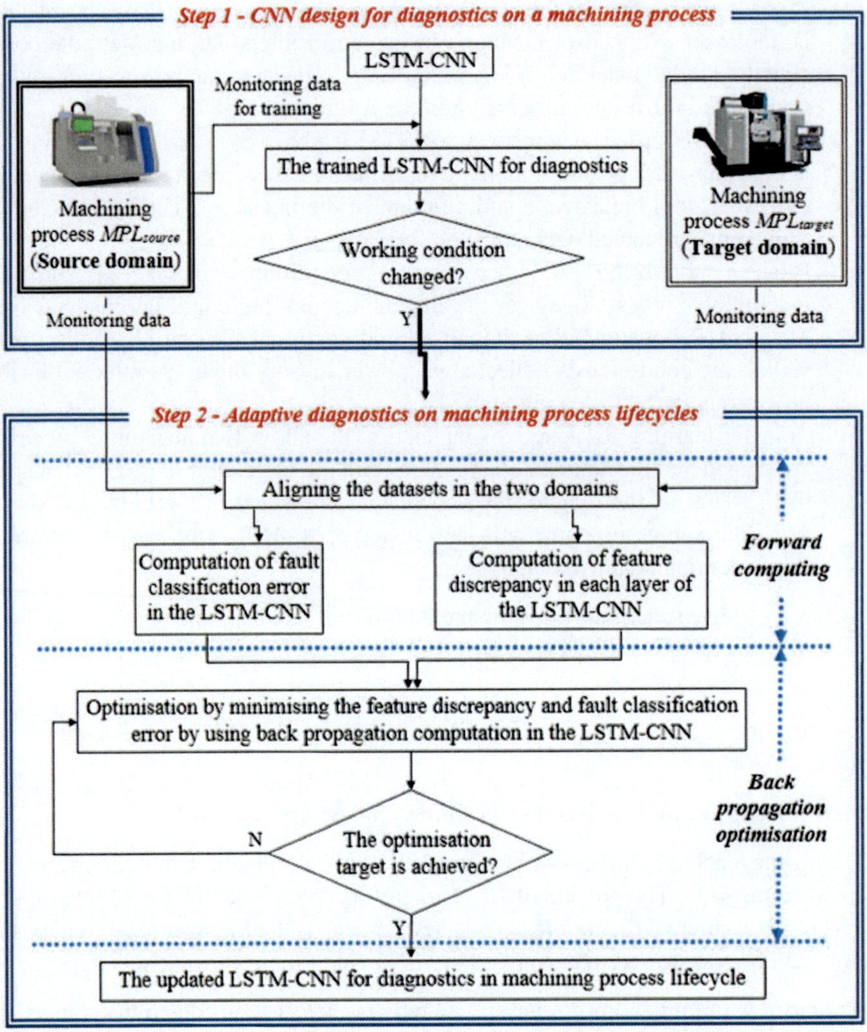

Fig. 2 Significant paradigm shifts of the global manufacturing industry

when there are further changes in working condition during the machining process lifecycle.

More technical details of the above major steps are depicted in the following Sects. 4and 5 respectively.

4 CNN Designed for Diagnostics on Machining Process

The CNN designed in this research consists of three primary layers, i.e., convolutional layer, max pooling layer and a fully connected layer. To improve the CNN, LSTM layer, Batch Normalization (BN), Rectified Linear Unit (ReLU) and Softmax are included.

One layer of LSTM is located at the beginning of the designed neural network, and the convolutional layer and max pooling layer are stacked twice for better performance. In LSTM layer, connection among data points can be built [31]. LSTM has the same topology as traditional Recurrent Neural Network (RNN), the accuracy of RNN is usually crippled by a vanishing/exploding gradients problem, LSTM can solve the problem by utilizing three gates: input gate, forget gate and output gate [32]. In the convolutional layer, neurons are connected based on a kernel filter with the same weights. In the max pooling layer, input values are down-sampled to extract core features. Finally, a fully connected layer is applied to classify the types of faults. The designed structure of the CNN is shown in Fig. 3. The input and output of the CNN are shown in Table 1. For a batch of machined work-pieces, the initial measured lengths of the signals are different. To facilitate the CNN processing, the maximum length among the input signals is chosen as the standard length. Shorter signals are

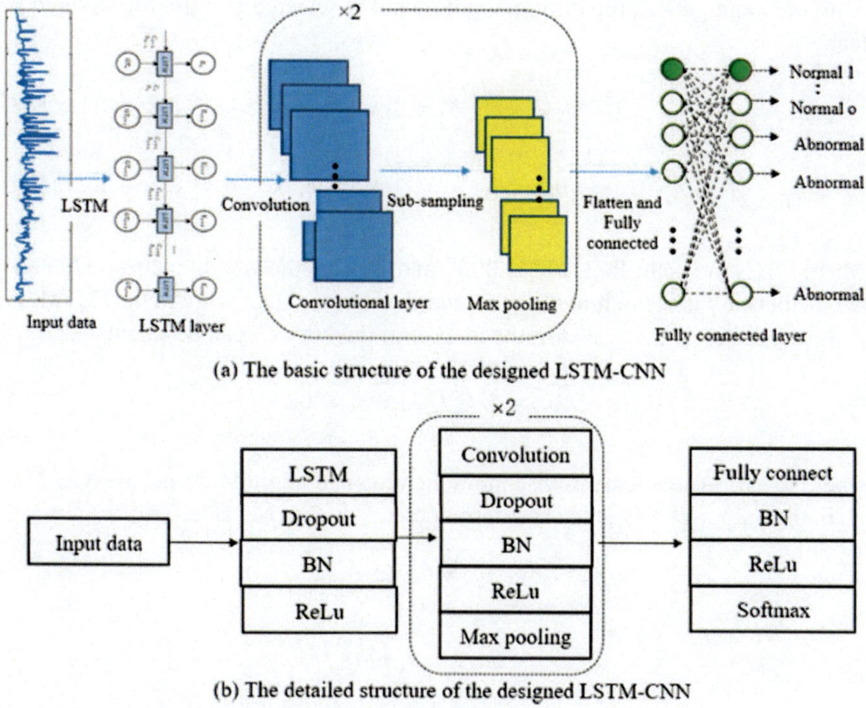

Fig. 3 The structure of the LSTM-CNN for diagnostics

Table 1 Input and output of the LSTM-CNN

Input vector (X)	Output vector (output)
Index *1* in power signal	Normal condition *1* [1, 0, 0, …, 0]
Index *2* in power signal	Abnormal condition *1* [0, 1, 0, …, 0]
… …	… …
Index *o* in power signal	Abnormal situation *o* [0, …, 1, …,0]
… …	… …
Index *m* in power signal	Abnormal situation *p* [0, 0, 0, …, 1]

zero padded to be in the standard length. The detailed explanation of the CNN design is given below.

In the forget gate f_t, part of values is removed by utilizing a sigmoid activation function:

$$f_t = \sigma\left(W_{GTf} \cdot X_t + W_{hf} \cdot h_{i-1} + b_f\right) \tag{4}$$

where σ is a sigmoid activation function; i is the counter for data; h_{i-1} is the short-term memory; W_{GTf} and W_{hf} are the weight matrices for X_i and h_{i-1}, respectively; X_i is the input defined in Table 1; b_f is the bias in the forget gate.

In the input gate i_i, the information to input is selected by utilizing sigmoid and tanh:

$$i_i = \sigma(W_{GTi} \cdot X_i + W_{hi} \cdot h_{i-1} + b_i) \tag{5}$$

$$\widetilde{C}_i = tanh(W_{GTc} \cdot X_i + W_{hc} \cdot h_{i-1} + b_c) \tag{6}$$

where tan is the tanh; W_{GTi}, W_{hi}, W_{GTc} and W_{hc} are the weight matrices; b_i and b_c are the bias. $\widetilde{C}_i$ is the output of the standard recurrent layer without the LSTM.

Based on C_{i-1}, $\widetilde{C}_i$, f_i and i_i, the long-term memory C_i can be calculated:

$$C_i = f_i \otimes C_{i-1} \oplus i_i \otimes \widetilde{C}_i \tag{7}$$

where $\otimes$ and $\oplus$ are element-wise multiplication and addition, respectively.

In the output gate O_i, the short-term memory h_t can be calculated:

$$O_i = \sigma(W_{GTo} \cdot X_i + W_{ho} \cdot h_{i-1} + b_o) \tag{8}$$

$$h_i = O_i \otimes tanh(C_i) \tag{9}$$

where W_{GTo} and W_{ho} are weight matrices; b_o is the bias.

In the convolutional layer, features are extracted from the input data through a kernel filter [33]:

$$\boldsymbol{O} = \boldsymbol{W}_c * \boldsymbol{h}_t + \boldsymbol{b}_c \tag{10}$$

where $\boldsymbol{O}$ is the output matrix after the convolutional layer; $*$ is the convolutional operator; $\boldsymbol{W}_c$ is the weight matrix of the kernel filter; $\boldsymbol{b}_c$ is the bias matrix.

In the convolutional layer, the following strategies are applied to enhance the performance of the LSTM-CNN:

(1) Batch Normalization (BN), which was firstly proposed by Ioffe and Szegedy [34] to reduce internal covariate shift during the training process. It is easier and quicker to train models with saturating nonlinearities. The values in the output O of the above convolutional layer are normalized via BN:

$$\mu_O = \frac{1}{n}\sum\nolimits_{i=1}^{n} O(i) \tag{11}$$

$$\sigma_O^2 = \frac{1}{n}\sum\nolimits_{i=1}^{n} (O(i) - \mu_X)^2 \tag{12}$$

$$\overline{O}(i) = \frac{O(i) - \mu_O}{\sqrt{\sigma_O^2 + \varepsilon}}, (i = 1, \ldots n) \tag{13}$$

where μ_O is the mean value of $O(i)$ ($O(i) \in \boldsymbol{O}$, n is the total number of the values in $\boldsymbol{O}$); σ_O is the standard deviation of $\boldsymbol{O}$; ε is a constant added to the batch variance for stability; $\overline{O}(i)$ is the normalized output.

(2) Rectified Linear Unit (ReLU), which is to avoid the CNN to be a linear model in dealing with non-linear problems [33]. It is an activation function as follows:

$$O'(i) = \max\left(0, \overline{O}(i)\right) \tag{14}$$

All the value $O'(i)$ are assembled into a matrix $\boldsymbol{O}'$. Then, a max pooling layer follows the convolutional layer to take a lower resolution of $\boldsymbol{O}'$ as follows:

$$\boldsymbol{O}'' = \max_{size(\gamma)=S}\left(\boldsymbol{O}'\right) \tag{15}$$

where $\boldsymbol{O}''$ is the output after the max pooling layer in max pooling blocks; S is the size of the max pooling blocks; The size of $\boldsymbol{O}''$ is reduced in comparison with the input matrix $\boldsymbol{O}'$, so that most significant features are extracted and kept in the matrix.

In this research, the above convolutional and max pooling layers are designed to repeat twice. The data will be processed in the convolutional layer and max pooling layer until reaching the fully connected layer. The computation in the fully-connected layer is below:

$$C = W_{full} * O'' + b_{full} \quad (16)$$

where C is the output; O'' are the values after the final max pooling layer; W_{full} and b_{full} are the weights and bias in the fully-connected layer, respectively.

The non-linear CNN has the overfitting issue due to complex co-adaptations [35]. In order to avoid complex co-adaptations on the training data, 'Dropout' has been proposed [36]. Each hidden unit is randomly omitted from the network with pre-defined probability (e.g. 0.2), so the overfitting effect on other hidden unit can be effectively reduced. BN and ReLu are also applied to optimize the performance of the CNN. Furthermore, to eliminate the adverse effect of extreme values without removing them from the dataset, Softmax is used to preserve the significance of data within a standard deviation of the mean. The transformed data C' can be calculated as follows [37]:

$$C(i)^{'} = \frac{1}{1 + e^{-\left(\frac{C(i)-\mu_C}{\sigma_C}\right)}} \quad (17)$$

where $C(i)^{'} \in C$, μ_C is the mean value of C; σ_C is the standard deviation of C.

For the above LSTM-CNN, it is used for diagnosing machining processes. The following three types of faults are identified with fault features (shown in Fig. 4):

- Severe tool wear: the level of power shifts vertically significantly, but the level of power during the idle stage remains the same [38].

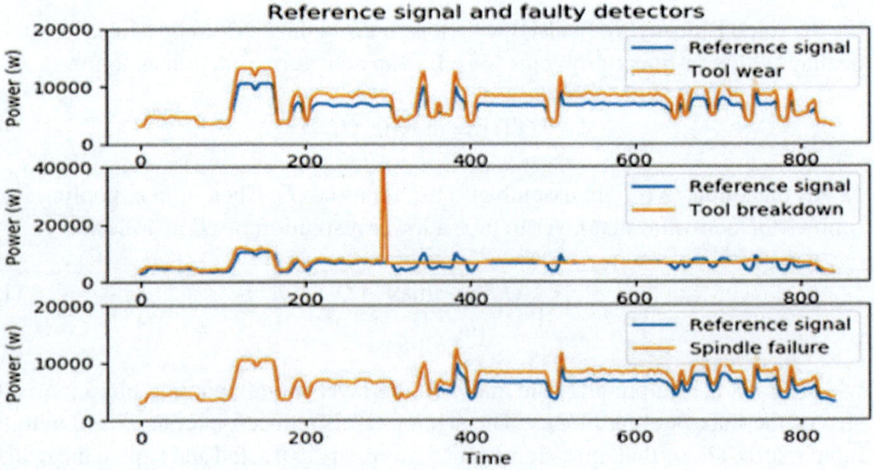

Fig. 4 Significant paradigm shifts of the global manufacturing industry

- Tool breakdown: power has a sudden peak and goes back to the air cutting level [38].
- Spindle failure: power has a sudden peak and an increased power level during both machining and idle stages [39].

5 Adaptive Diagnostics on Machining Process Lifecycle

For a new machining condition, it usually requires a sufficient set of labeled data for re-training the LSTM-CNN to ensure the high accuracy of diagnostics. However, this requirement is impractical during machining process lifecycles due to the high cost and long time for re-collecting and re-labeling the dataset during new conditions for LSTM-CNN re-training. LSTM-CNN is offline procedure with historical data processing.

To address the challenge, in this research, a transfer learning strategy is designed and integrated with the above LSTM-CNN as novel adaptive diagnostics in supporting machining process lifecycles. The strategy consists of two steps: (1) Forward computing of the LSTM-CNN using datasets optimally aligned from different domains; (2) Back propagation optimization on the LSTM-CNN to minimize domain-based feature distribution discrepancies as well as fault classification error by re-weighting the LSTM-CNN. The strategy is illustrated in Fig. 5 and detailed in the following sub-sections.

5.1 Forward Computing Using Optimally Aligned Data

For the same work-piece machined using the same process parameters but under different machines, signal patterns should be similar. Thus, the trained LSTM-CNN

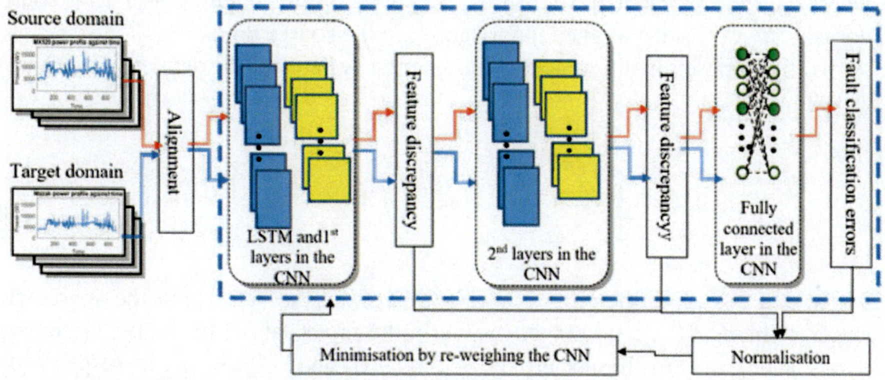

Fig. 5 The framework of deep transfer learning for adaptive diagnostics

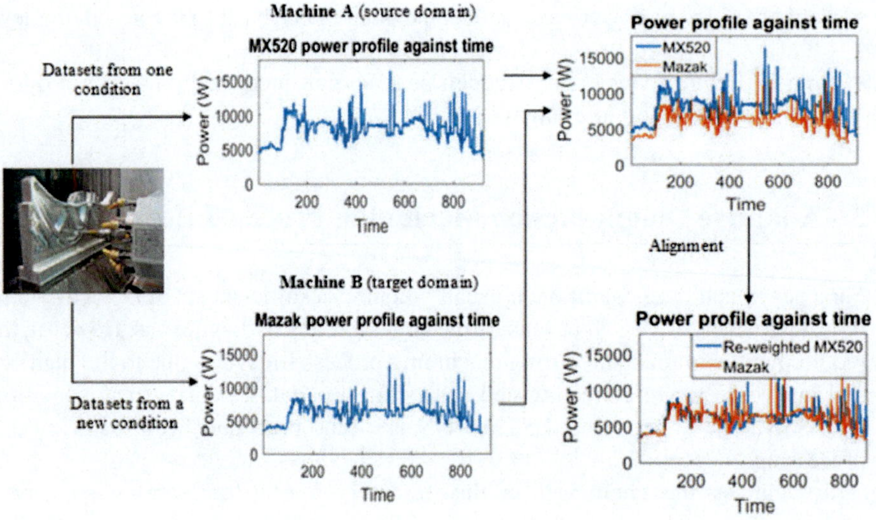

Fig. 6 Alignment on the datasets from the two domains

for the source domain (e.g., MP_i under Machine M_i) could be considered for the target domain (e.g., MP_{i_r} under Machine M_{i_r}). However, there are still magnitude differences between the input datasets from the different domains. Therefore, alignment optimization on the signals is required to effectively implement cross-domain knowledge transfer. The alignment process is illustrated in Fig. 6, and the optimization process is detailed below.

The data in the source domain for a work-piece can be adjusted for alignment as the following:

$$X'_{source} = \alpha \cdot X_{source} \tag{18}$$

where α is the alignment coefficient; X_{source} is the original input dataset in the source domain; X'_{source} is the aligned input dataset in the source domain.

To align datasets in different domains, similarity between the datasets is calculated as follows:

$$\sigma_{target\_source} = \frac{1}{m-1} \cdot \sum\nolimits_{i=1}^{m} \left(\left(X_{target}(i) - \mu_{target} \right) \cdot \left(X'_{source}(i) - \mu'_{source} \right) \right) \tag{19}$$

where σ_{taget_source} is the cross-covariance between the datasets in the source and target domains; $X_{target}(i)$ is No. i value in the input dataset of the target domain; μ_{target} and μ'_{source} are the means of $\sum_{i=1}^{m} X_{target}(i)$ and $\sum_{i=1}^{m} X'_{source}(i)$ respectively; m is the maximum number of the datasets. α in Eq. (12) can be determined through the following optimization process:

$$Maximize(\sigma_{target\_source})\text{subjectto}\alpha \in [0, u]$$

$$u = ceil\left(\frac{\mu_{target}}{\mu_{source}}\right) \tag{20}$$

where α is searched within a constrained scope [0, u]; ceil is the function to round the value to the next bigger integer.

The pattern search algorithm is applied to find the optimum of α as it does not require the gradient of the objective function.

[Optimization Procedure I]—Optimization of the alignment coefficient α.

1. Choose an initial point a_0 and define pattern vectors. For a problem with a single input variable, there are two pattern vectors: $v_1 = [1]$ and $v_2 = [-1]$;
2. Search for a mesh point α_i around a_0 that has a smaller objective function compared to a_0. The search mesh is generated as $\alpha_i = a_0 + \Delta_i$, where $\Delta_i = \Delta m \cdot v_i$, and Δm is the current mesh size;
3. If a better solution α_i is found, update $a_0 = \alpha_i$ and increase the mesh size: $\Delta m = 2 \cdot \Delta m$, otherwise, keep the original a_0 and reduce the mesh size: $\Delta m = 0.5 \cdot \Delta m$;
4. Check if any of the stop conditions is met (the maximum epoch number is reached or the algorithm converges). If yes, stop the optimization. Otherwise, go to above Step 2.

5.2 *Back Propagation Optimization for Minimizing Feature Distribution Discrepancy and Classification Error*

To transfer the knowledge from the source domain to the target domain, the trained LSTM-CNN should be subject to the condition of that features are in similar distributions between the domains. To address feature distribution mismatch during transfer learning, an optimization strategy is applied through back propagation computing on the designed LSTM-CNN.

In the previous literature, MMD (Maximum Mean Discrepancy) was popularly used to measure the distance metric for distribution probabilities between two domains. That is, the datasets in the source domain and the target domain are represented as $D_{source} = \{X_{source}, P(X_{source})\}$ and $D_{target} = \{X_{target}, P(X_{target})\}$ respectively (D_{source}, D_{target} are the source and target domains; X_{source}, X_{target} are the input datasets of the source and target domains; $P(X_{source})$, $P(X_{target})$ are the distribution probabilities of the source and target domains). The MMDs of the source and target domains are defined below:

$$Mean_H(\boldsymbol{X_{source}}) = \frac{1}{m}\sum\nolimits_{i=1}^{m} H(X_{source}(i)) \tag{21}$$

$$Mean_H(\boldsymbol{X}_{target}) = \frac{1}{m}\sum\nolimits_{j=1}^{m} H(X_{target}(i)) \quad (22)$$

$$MMD_H(\boldsymbol{X}_{source}, \boldsymbol{X}_{target}) = sup(Mean_H(X_{source}) - Mean_H(X_{target})) \quad (23)$$

where $sup(\cdot)$ is the supremum of the aggregate; $H(\cdot)$ is a RKHS (Reproducing Kernel Hilbert Space).

In this research, the concept of MMD is adopted for measuring the distribution difference of domain invariant features. In the LSTM-CNN design presented earlier, there are two repeated convolutional and max pooling layers, and $MMD_{H1}(\boldsymbol{X}_{source}, \boldsymbol{X}_{target})$ and $MMD_{H2}(\boldsymbol{X}_{source}, \boldsymbol{X}_{target})$ are used to measure the feature distribution distance for different domains in these two layers respectively. To achieve the similar distributions from two domains, $MMD_{H1}(\boldsymbol{X}_{source}, \boldsymbol{X}_{target})$ and $MMD_{H2}(\boldsymbol{X}_{source}, \boldsymbol{X}_{target})$ will be considered as the optimization objectives to regularize the weights of the LSTM-CNN. To justify the usage of both $MMD_{H1}(\boldsymbol{X}_{source}, \boldsymbol{X}_{target})$ and $MMD_{H2}(\boldsymbol{X}_{source}, \boldsymbol{X}_{target})$, a benchmark has been done for comparison in Sect. 6.3. Meanwhile, during the re-weighting process on the LSTM-CNN, the classification error between labeled data and predicted data in should be minimized as well. Thus, the classification error is considered as the third optimization objective.

The cross entropy $cross_{entropy}$ is used to indicate the classification error, which is calculated below:

$$cross_{entropy} = -\frac{1}{k}\left|\sum\nolimits_{i=1}^{k} (\hat{y}lny + (1 - \hat{y})\ln(1 - y))\right| \quad (24)$$

where y is the predicted output; $\hat{y}$ is the true output.

As aforementioned, the total loss function $Loss$ can be modelled based on the above three objectives, i.e.,$MMD_{H1}(\boldsymbol{X}_{source}, \boldsymbol{X}_{target})$, $MMD_{H2}(\boldsymbol{X}_{source}, \boldsymbol{X}_{target})$ and $cross_{entropy}$. Since they are in different value ranges, normalization is required before optimization.

In this research, Nadir and Utopia points are utilized to normalize the above three objectives in an effective means [40]. The Utopia point z_i^U provides the lower bound of No. i objective obtained by minimizing the objective as below:

$$z_i^U = minf(i) \quad (25)$$

The Nadir point z_i^N provides the upper bound of No. i objective by maximizing the objectives:

$$z_i^N = \max_{1 \ll j < I} f(j) \quad (26)$$

where I is the total number of the objective functions.

According to Eqs. (19) and (20), the normalized $MMD_{H1}(X_{source}, X_{target})$, $MMD_{H2}(X_{source}, X_{target})$ and $cross_{entropy}$ can be calculated below (in this case, MMD_{H1} is normalized when $i = 1$, MMD_{H2} is normalized when $i = 2$, $cross_{entropy}$ is normalized when $i = 3$):

$$nMMD_{H1} = \left(MMD_{H1}\left(\boldsymbol{X}_{\boldsymbol{source}}, \boldsymbol{X}_{\boldsymbol{target}}\right) - z_1^u\right) / \left(z_1^N - z_1^u\right) \tag{27}$$

$$nMMD_{H2} = \left(MMD_{H2}\left(\boldsymbol{X}_{\boldsymbol{source}}, \boldsymbol{X}_{\boldsymbol{target}}\right) - z_2^u\right) / \left(z_2^N - z_2^u\right) \tag{28}$$

$$ncorss = \left(cross_{entropy} - z_3^u\right) / \left(z_3^N - z_3^u\right) \tag{29}$$

where $nMMD_{H1}$, $nMMD_{H2}$ and ncorss are the normalized $MMD_{H1}\left(\boldsymbol{X}_{\boldsymbol{source}}, \boldsymbol{X}_{\boldsymbol{target}}\right)$, $MMD_{H2}\left(\boldsymbol{X}_{\boldsymbol{source}}, \boldsymbol{X}_{\boldsymbol{target}}\right)$ and cross_entropy.

To avoid overfitting, L2 Regularization is also included. The total loss function Loss can be calculated based on the weighted sum of the three normalized objectives [41]:

$$Loss = w_1 \cdot nMMD_{H1} + w_2 \cdot nMMD_{H2} + w_3 \cdot ncorss + w_4 \cdot W + w_5 \cdot accuracy \tag{30}$$

$$accuracy = \frac{N}{T} \times 100\% \tag{31}$$

where $w_1 - w_5$ are the weights for the three objectives respectively, and $\sum_{i=1}^{5} w_i = 1$, W is the matrix of weights in designed LSTM-CNN; N the number of correct classification and T is total number of samples.

The total loss function is minimized using an Adam optimization process during the back propagation computing process of the LSTM-CNN as the following:

[Optimization Procedure II]—Back propagation optimization on the LSTM-CNN.

1. Define $\boldsymbol{W}_{\boldsymbol{LSTM-CNN}}$ and $\boldsymbol{b}_{\boldsymbol{LSTM-CNN}}$ as weight and bias matrices of the LSTM-CNN, which are initialized with random values;
2. Set the maximum epoch time E;
3. Update $\boldsymbol{W}_{\boldsymbol{LSTM-CNN}}$ by minimizing the total loss function with Adam:

$$\boldsymbol{W}^{t+1}_{\boldsymbol{LSTM-CNN}} = \boldsymbol{W}^{t}_{\boldsymbol{LSTM-CNN}} - \varepsilon \cdot \left(\frac{\partial Loss}{\partial \|\boldsymbol{W}_{\boldsymbol{LSTM-CNN}}\|}\right) \tag{32}$$

where t is the current epoch number; ε is the learning rate, which can be calculated below [42]:

$$\varepsilon = \varepsilon_0 \cdot m_t / \sqrt{v_t} \tag{33}$$

$$m_t = \beta_1 \cdot m_{t-1} + (1 - \beta_1) \cdot g_t \quad (34)$$

$$v_t = \beta_2 \cdot v_{t-1} + (1 - \beta_2) \cdot g_t^2 \quad (35)$$

where ε_0 is the initial learning rate; β_1 and β_2 are the hyper-parameters and g_t is the current gradient; m_t and v_t are the moments.

4. Update $\boldsymbol{b}_{LSTM-CNN}$ by minimizing the total loss function with Adam:

$$\boldsymbol{b}_{LSTM-CNN}^{t+1} = \boldsymbol{b}_{LSTM-CNN}^{t} - \varepsilon \cdot \left(\frac{\partial Loss}{\partial ||\boldsymbol{b}_{LSTM-CNN}||} \right) \quad (36)$$

5. Repeat the above Steps 3 and 4 until the maximum epoch time is reached or the algorithm converges.

The well-trained LSTM-CNN model with sufficient data can be used for failure diagnosis for the particular machine/tooling. The parameters of the LSTM-CNN can be transferred to that of a new machine condition, as the parameters of trained CNN model are usually closer to the optimum parameters than that of initialized parameters. The LSTM-CNN can then be fine-tuned based on power data patterns. Therefore, the training workload and time can be significantly reduced. In order to justify the advantage of parameter transfer, training performance has been compared for benchmark in Sect. 6.3.

6 Experiments and Discussions

6.1 Experiment Set-Up

The data for the developed approach were collected in a machining company in the UK. The company specializes on mass-customized and high-precision machining products, which include chuck jaws, clamping, tooling accessory, etc. To verify the approach presented in this chapter, machining conditions based on two CNC machines (MX520 and Mazak VTC-800/20SR machine) were monitored. The collected power data were transmitted to a data server for processing via a gateway and the MQTT communication protocol.

During the machining processes, the data used to train the CNN were collected from the MX520 machine for six months (the source domain) [30]. The Mazak VTC-800/20SR machine was used as an alternative machine during the machining lifecycle when the MX520 was replaced (the target domain).

Table 2 Work-pieces (WP-) and working conditions (C-) in both domains

Target domain	WP-1	WP-2	WP-3	WP-4	WP-5
Normal (C-1)	20	17	18	21	17
Tool wear (C-2)	1	0	1	1	1
Tool Breakage (C-3)	0	1	0	1	0
Spindle failure (C-4)	0	0	0	1	0
Total samples: 100; There are 20 classes (5 componentsx4 conditions)					
Source domain	WP-1	WP-2	WP-3	WP-4	WP-5
Normal (C-1)	1420	291	372	393	572
Tool wear (C-2)	31	21	23	12	20
Tool Breakage (C-3)	11	9	3	4	5
Spindle failure (C-4)	3	3	2	2	3
Total samples: 3200; There are 20 classes (5 componentsx4 conditions)					

The data collected for the two machines are summarized in Table 2. In the data, there are five types of work-pieces machined by the two machines and four conditions (one normal and three abnormal). The LSTM-CNN was trained, validated and tested using 3200 samples collected on MX520 (i.e., the source domain). Among the 3200 samples, 70% were used for training, 15% for validation and 15% for testing of the LSTM-CNN. The transfer learning mechanism was applied on Mazak VTC-800/20SR to implement domain adaption for the trained LSTM-CNN. In this case study, the ratio of datasets between the target and source domains is 3.125%.

6.2 *Optimized Alignment of the Datasets Between Domains*

To use the data of the source domain (MX520) and the target domain (Mazak VTC-800/20SR) to implement transfer learning, the magnitude difference of the datasets between the domains was optimally aligned. For the same work-piece machined by the two equipment using the same process parameters, the alignment coefficient α can be determined by the Optimization Procedure I defined in Sect. 5.1. For instance, in machining work-piece 1, α was identified as 0.81. Thus, $X^{'}_{source}$ can be calculated as $0.81 \cdot X_{source}$ to minimize the dataset differences of the two domains. In machining work-piece 2, α was identified as 0.83, so that $X^{'}_{source}$ can be calculated as 0.83 · X_source accordingly. Figure 7 shows a product assembly (scroll jaw) and its two work-pieces for comparison here. Figure 8 shows the aligned examples of the two work-pieces machined on MX520 and Mazak VTC-800/20SR.

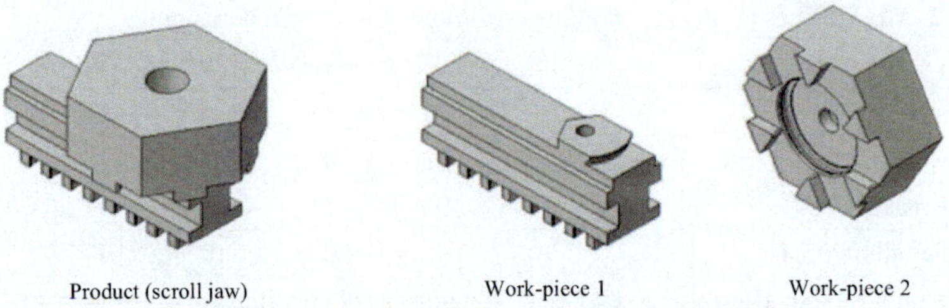

Fig. 7 A product assembly and its two work-pieces

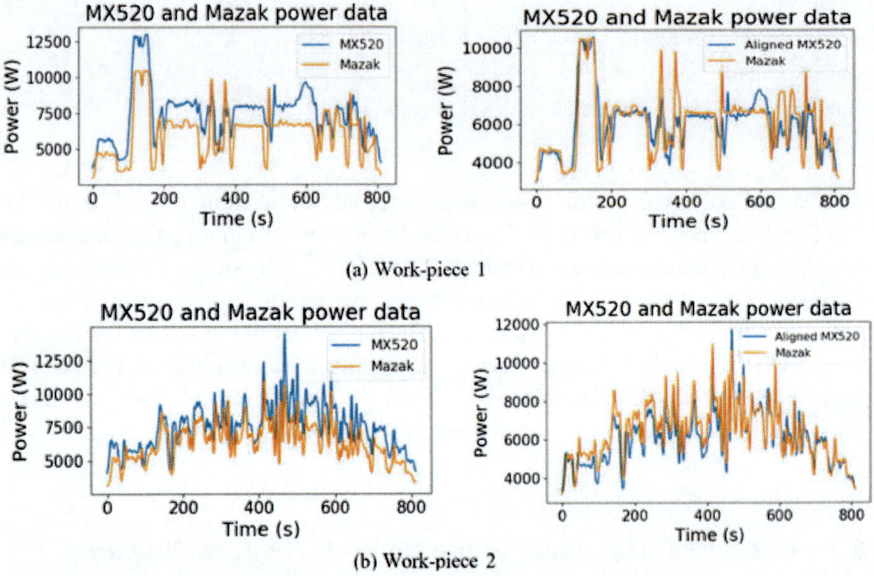

Fig. 8 Aligned data for two work-pieces machined using MX520 and Mazak (the left is the original data, and the right is the aligned data)

6.3 *Minimization of Feature Distribution Discrepancy Between Domains and Classification Error*

For the working conditions from the two machines for manufacturing the same work-piece under the same process, the fault features are similar but their distributions are different. Therefore, the Optimization Procedure II defined in Sect. 5.2 was applied. Here, the two MMDs and entropy were considered equally important, so that w_1, w_2, w_3, w_4 and w_5 were set as 1/5 each in Eq. (30).

For the MMD in a RKHS, different training approaches, including Gaussian kernel with/without aligned input data, and Radial Basis Function (RBF) with/without aligned input data, were compared over 100 epochs (results are shown in Fig. 9).

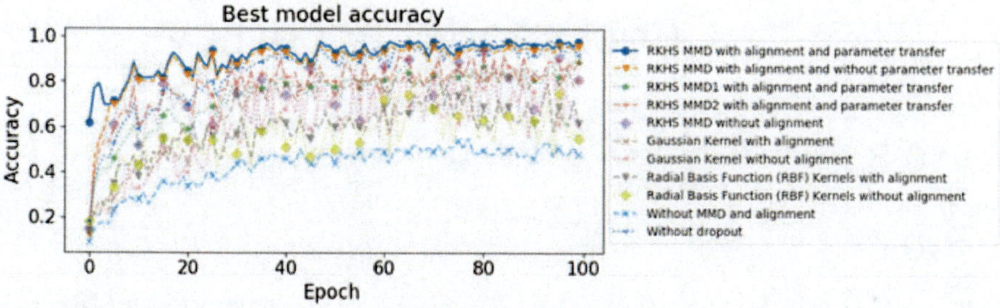

Fig. 9 The achieved best accuracy for different approaches

Figure 10 and Table 3 summarize the training/validation accuracy for each used kernel and setting.

It indicates that the MMD in a RKHS with data alignment and parameter transfer achieved the best training accuracy (all the accuracy is calculated based on Eq. (31)): the best accuracy (98%/95%), the worst accuracy (90%/87%) and the average accuracy (95%/91%). Meanwhile, comparisons were made to indicate that the accuracy of the developed approach is much higher than the approaches without MMD and data alignment.

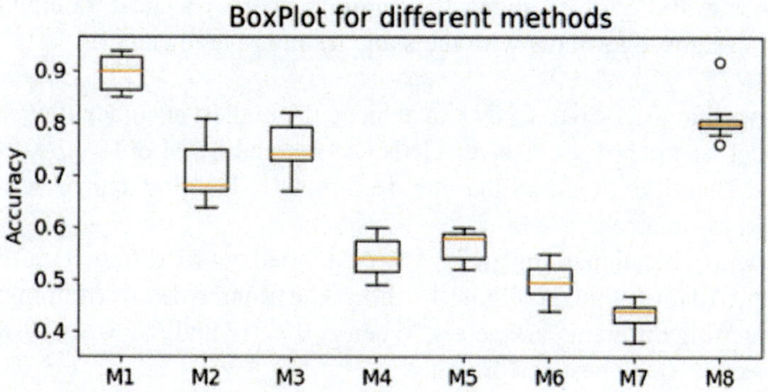

Fig. 10 BoxPlot for the different approaches

Table 3 Benchmark for different approaches with different settings

	Best accuracy	Worst accuracy	Average accuracy
LSTM-CNN	0.98/0.95	0.90/0.87	0.95/0.91
2-layer CNN	0.94/0.90	0.85/0.81	0.90/0.85
1-layer CNN	0.86/0.83	0.70/0.69	0.78/0.73
ANNs	0.83/0.79	0.67/0.61	0.74/0.70
SVM	0.60/0.58	0.48/0.43	0.53/0.50

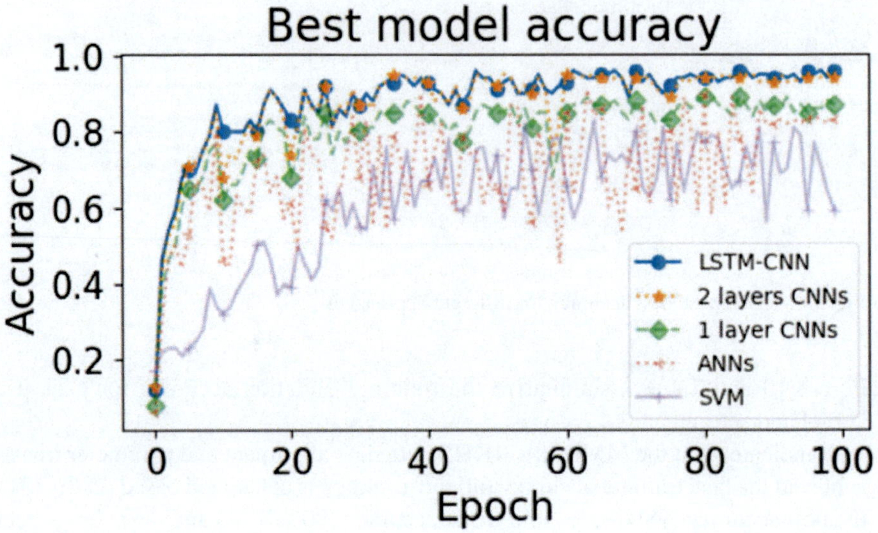

Fig. 11 Accuracy for different CNN designs and algorithms

For the developed LSTM-CNN, different designs were compared for benchmarking as well. Figure 11 shows the training accuracy against training epochs for different designs/algorithms with the same training parameters ($\varepsilon_0 = 0.1$, $\beta_1, \beta_2 = 0.9$).

It shows that the 2-layer CNN can achieve the highest accuracy (96%), while the best model accuracy for a 1-layer CNN, ANNs and SVM did not exceed 90% in accuracy. Therefore, it shows that the deep transfer learning approach has the best training performance.

Meanwhile, benchmarking analyses were carried out for different training parameters in the Adam optimizer. Figure 12 shows the training details (training/validation accuracy) using different parameters. When $\varepsilon\_0 = 0.1$ and $\beta_1, \beta_2 = 0.9$, the training process achieved the best performance.

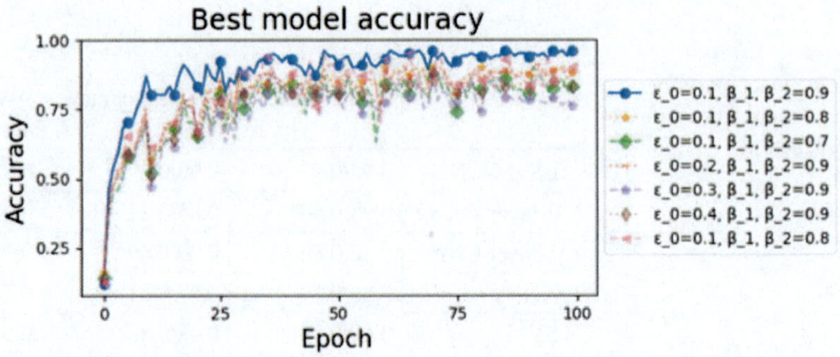

Fig. 12 The training details using different parameters

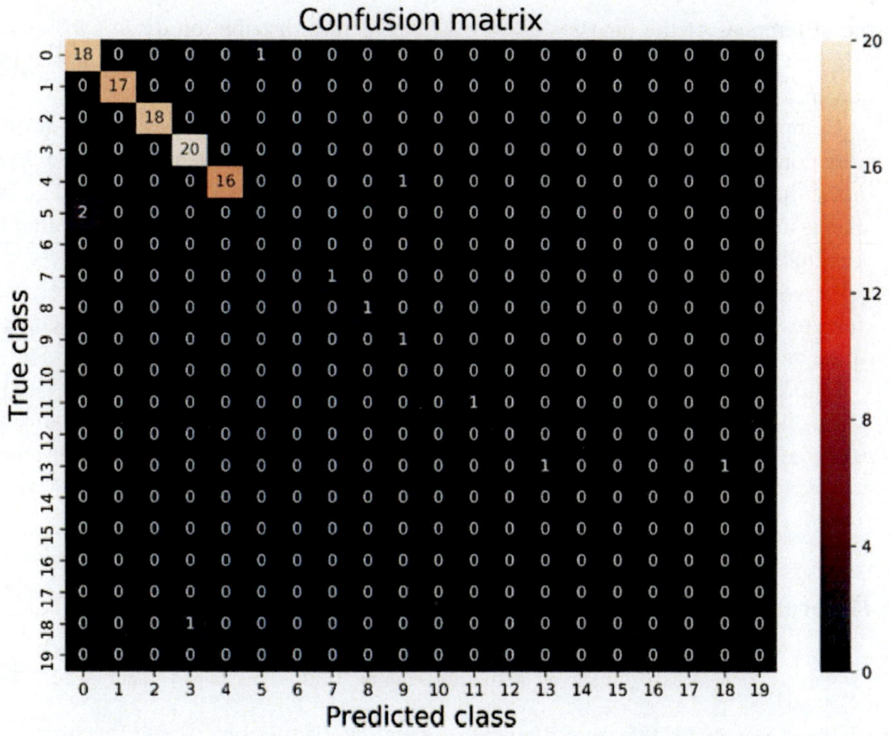

Fig. 13 The confusion matrix for the 2-layer LSTM-CNN for target domain

Figure 13 shows the confusion matrix of the LSTM-CNN for the best accuracy result in the target domain. It can be noticed that in the matrix, some classes do not have specific sample for testing (e.g., label 6). The reason is that, as aforementioned, it is difficult to collect a large number of faulty patterns in the target domain so that the transfer learning mechanism is developed in this research.

7 Conclusions

To achieve manufacturing sustainability, it is imperative to develop an adaptive diagnostics approach to minimize faults throughout machining process lifecycles. With this aim, this chapter presents a new deep transfer learning approach. In the approach, a LSTM-CNN is designed to conduct diagnostics on machining processes. A deep learning strategy is further incorporated into the LSTM-CNN to enhance the adaptability of the approach during machining lifecycles. The innovations are from the following aspects: (1) It is a new attempt to design a new deep transfer learning approach to perform adaptive diagnostics on machining process lifecycles; (2) It is also innovative in new optimization mechanisms for transfer learning to minimize the differences of the datasets as well as the feature distribution discrepancies on different machining conditions. Meanwhile, parameter transfer is also proposed for new machine conditions to enhance the training accuracy and efficiency.

Comparative evaluations on various parameters, settings and algorithm design were conducted showing that the approach presented in this chapter achieved 96% in accuracy, which is significantly higher than other approaches. Meanwhile, case studies in the research showed that failure modes under different operating conditions are almost impossible to obtain in practices. Data for an exemplary machine were collected for 6 months with different faulty patterns, so that those faulty patterns (features) can be kept and transferred to new machines to ensure the training accuracy of the developed system.

Future research works include how to evaluate the knowledge from multiple source domains and furthermore transfer them to new domains. The research is expected to further enhance the robustness and reliability of the approach for more complex manufacturing systems.

References

1. Pan SJ, Yang Q (2010) A survey on transfer learning. IEEE Trans Knowl Data Eng 22:1345–1359
2. Jiang Y et al (2017) Seizure classification from EEG signals using transfer learning, semi-supervised learning and TSK fuzzy system. IEEE Trans Neural Syst Rehabil Eng 25:2270–2284
3. Chen G, Li Y, Liu X (2019) Pose-dependent tool tip dynamics prediction using transfer learning. Int J Mach Tools Manuf 137:30–41
4. Talo M, Baloglu U, Yıldırım Ö, Rajendra AU (2019) Application of deep transfer learning for automated brain abnormality classification using MR images. Cognitive Syst Res 54:176–188
5. Yin X, Li Z (2016) Reliable decentralized fault prognosis of discrete-event systems. IEEE Trans Syst, Man, Cybern: Syst 46(11):1598–1603
6. Lu S, He Q, Yuan T, Kong F (2016) Online fault diagnosis of motor bearing via stochastic-resonance-based adaptive filter in an embedded system. IEEE Trans Syst Man, Cybern: Syst 47:1111–1122
7. Wang T, Han Q, Chu F, Feng Z (2019) Vibration based condition monitoring and fault diagnosis of wind turbine planetary gearbox: A review. Mech Syst Signal Process 126:662–685
8. Tian J, Morillo C, Azarian M, Pecht M (2016) Motor bearing fault detection using spectral kurtosis-based feature extraction coupled with k-Nearest Neighbor distance analysis. IEEE Trans Industr Electron 63:1793–1803
9. Zhou Z, Wen C, Yang C (2016) Fault isolation based on k-Nearest Neighbor rule for industrial processes. IEEE Trans Industr Electron 63:1–1
10. Li F, Wang J, Tang B, Tian D (2014) Life grade recognition method based on supervised uncorrelated orthogonal locality preserving projection and K-nearest neighbor classifier. Neurocomputing 138:271–282
11. Han H, Cui X, Fan Y, Qing H (2019) Least squares support vector machine (LS-SVM)-based chiller fault diagnosis using fault indicative features. Appl Therm Eng 154:540–547
12. Suo M, Zhu B, An R, Sun H, Xu S, Yu Z (2019) Data-driven fault diagnosis of satellite power system using fuzzy Bayes risk and SVM. Aerosp Sci Technol 84:1092–1105
13. Luo B, Wang H, Liu H, Li B, Peng F (2019) Early fault detection of machine tools based on deep learning and dynamic identification. IEEE Trans Industr Electron 66:509–518

14. Zhong S, Fu S, Lin L (2019) A novel gas turbine fault diagnosis method based on transfer learning with CNN XE "CNN" . Measurement 137:435–453
15. Li K, Xiong M, Li F, Su L, Wu J (2019) A novel fault diagnosis algorithm for rotating machinery based on a sparsity and neighborhood preserving deep extreme learning machine. Neurocomputing 350:261–270
16. Afridi M, Ross A, Shapiro E (2018) On automated source selection for transfer learning in convolutional neural networks. Pattern Recogn 73:65–75
17. Yao Y, Doretto G (2010) Boosting for transfer learning with multiple sources. Proceedings of the 2010 IEEE computer society conference on computer vision and pattern recognition
18. Asgarian A, Sobhani P, Zhang JC, Mihailescu M, Sibilia A, Ashraf AB, Taati B (2018) Hybrid instance-based transfer learning method. Proceedings of the machine learning for health (ML4H) workshop at neural information processing systems
19. Feuz KD, Cook DJ (2015) Transfer learning across feature-rich heterogeneous feature spaces via feature-space remapping (FSR). ACM transactions on intelligent systems and technology (TIST). 6
20. Rozantsev A, Salzmann M, Fua P (2018) Beyond sharing weights for deep domain adaptation. IEEE Trans Pattern Anal Mach Intell 41:801–814
21. Yang Z, Dhingra B, He K, Cohen WW, Salakhutdinov R, LeCun Y (2018) Glomo: Unsupervisedly learned relational graphs as transferable representations. Available: arXiv preprint arXiv:1806.05662
22. Zhang L, Yang J, Zhang D (2017) Domain class consistency based transfer learning for image classification across domains. Inf Sci 418–419:242–257
23. Kim D, MacKinnon T (2018) Artificial intelligence in fracture detection: transfer learning from deep convolutional neural networks. Clin Radiol 73:439–445
24. Lu W, Liang B, Cheng Y, Meng D, Yang J, Zhang T (2017) Deep modelbased domain adaptation for fault diagnosis. IEEE Trans Industr Electron 64:2296–2305
25. Xiao D, Huang Y, Zhao L, Qin C, Shi H, Liu C (2019) Domain adaptive motor fault diagnosis using deep transfer learning. IEEE Access 7:80937–80949
26. Wen L, Gao L, Li X (2019) A new deep transfer learning based on sparse auto-encoder for fault diagnosis. IEEE Trans Syst, Man, Cybern: Syst 49:136–144
27. Sun C, Ma M, Zhao Z, Tian S, Yan R, Chen X (2019) Deep transfer learning based on Sparse Autoencoder for remaining useful life prediction of tool in manufacturing. IEEE Trans Industr Inf 15:2416–2425
28. Liu Z, Guo Y, Sealy M, Liu Z (2016) Energy consumption and process sustainability of hard milling with tool wear progression. J Mater Process Technol 229:305–312
29. Sealy M, Liu Z, Zhang D, Guo Y, Liu Z (2016) Energy consumption and modeling in precision hard milling. J Clean Product 135:1591–1601
30. Liang YC, Lu X, Li WD, Wang S (2018) Cyber Physical System and Big Data enabled energy efficient machining optimization. J Clean Product 187:46–62
31. Song X et al (2020) Time-series well performance prediction based on Long Short-Term Memory XE "Long Short-Term Memory" (LSTM XE "LSTM") neural network model. J Petrol Sci Eng 186:106682
32. Wang K, Qi X, Liu H (2019) Photovoltaic power forecasting based LSTM XE "LSTM" - Convolutional Network. Energy 189:116225
33. Jing L, Zhao M, Li P, Xu X (2017) A convolutional neural network based feature learning and fault diagnosis method for the condition monitoring of gearbox. Measurement 111:1–10
34. Ioffe S, Szegedy C (2015) Batch normalization: Accelerating deep network training by reducing internal covariate shift", Available: arXiv:1502.03167
35. Poernomo A, Kang D (2018) Biased dropout and crossmap dropout: Learning towards effective dropout regularization in convolutional neural network. Neural Netw 104:60–67
36. Hinton G, Srivastava N, Krizhevsky A, Sutskever I, Salakhutdinov R (2012) Improving neural networks by preventing co-adaptation of feature detectors. arXiv:1207.0580v1, 2012
37. Priddy K, Keller P (2005) Artificial neural networks. SPIE Press, Bellingham, Wash

38. Teti R, Jemielniak K, O'Donnell G, Dornfeld D (2010) Advanced monitoring of machining operations. CIRP Ann 59:717–739
39. Axinte D, Gindy N (2004) Assessment of the effectiveness of a spindle power signal for tool condition monitoring in machining processes. Int J Prod Res 42:2679–2691
40. Mausser H., 2019. Normalization and other topics in multi-objective optimization. Proceedings of the fields–MITACS industrial problems workshop. pp 59–101
41. Zeng S, Gou J, Deng L (2017) An antinoise sparse representation method for robust face recognition via joint l 1 and l 2 regularization. Expert Syst Appl 8(1):1–9
42. Kingma D, Ba J (2015) ADAM: A method for stochastic optimization", Available at arXiv: 1412.6980v9
43. Xu G, Liu M, Jiang Z, Shen W, Huang C (2020) Online fault diagnosis method based on transfer convolutional neural networks. IEEE Trans Instrum Meas 69:509–520

CNN-LSTM Enabled Prediction of Remaining Useful Life of Cutting Tool

X. Y. Zhang, X. Lu, W. D. Li, and S. Wang

1 Introduction

The condition of a cutting tool is considered as one of the most paramount factors to determine production quality, productivity and energy consumption [1]. If a computerized numerical control (CNC) machine is equipped with a system for tool condition monitoring and analysis to estimate the remaining useful life (RUL) of a cutting tool, it can prevent the insufficient utilization of the tool or the poor surface quality/dimensional errors of workpieces due to the excessive wear of the tool. Investigations showed that 30% of the manufacturing cost could be saved by adopting such the system [2]. Owing to the significance of the research, numerical works have been actively conducted in recent years.

The research of the RUL prediction on cutting tools can be classified into two general categories, i.e., an experience-based approach and a data-driven approach [3]. An experience-based approach requires a large amount of machining cycles to be carried out to summarize laws of tool wear as mathematical models or some heuristic rules (e.g., the Taylor law). However, the status of a cutting tool is significantly influenced by various machining parameters, such as cutting speed and feed rate. Furthermore, as a cutting tool is constantly contacted with a workpiece during machining, this process unavoidably involves many material-specific factors that stochastically affect the tool condition. Thus, it is difficult to generalize an experience-based approach for accurately predicting the RUL of a cutting tool in wider applications. Instead, a data-driven approach, which estimates the RUL of a cutting tool based on analysis of monitored signals from sensors mounted on a machine or a

X. Y. Zhang · X. Lu · W. D. Li (✉) · S. Wang
Faculty of Engineering, Environment and Computing, Coventry University, Coventry, UK
e-mail: weidong.li@coventry.ac.uk

W. D. Li
School of Logistics Engineering, Wuhan University of Technology, Wuhan, China

W. Li et al. (eds.), *Data Driven Smart Manufacturing Technologies and Applications*, Springer Series in Advanced Manufacturing,
https://doi.org/10.1007/978-3-030-66849-5_5

cutting tool, shows a greater potential for practical applications. The earlier design of such the approach depends on support vector machine (SVM) or support vector regression (SVR) [4], artificial neural networks (ANN) [5], fuzzy logic [6], etc., for analyzing and estimating tool conditions and lives. However, those "conventional" intelligent algorithms still need tedious manual operations to define features for extraction from monitored signals, and the pre-set features might not be consistent throughout an actual machining process. Another disadvantage is that those algorithms are incapable of handling a large volume of sensor data captured during the entire lifecycle of a machining process. In recent years, deep learning algorithms, such as the convolutional neural network (CNN), recurrent neural network (RNN) and long short-term memory network (LSTM), have been increasingly used to address the above challenges. The deep learning algorithms can facilitate a data-driven approach in extracting features from a large amount of monitored datasets more intelligently and autonomously. Nevertheless, industrial applications have been becoming more complex, requiring a more accurate and efficient approach. Therefore, there are some further exploration areas, including the followings:

Each deep learning algorithm exhibits distinguishing characteristics but shows some limits. For processing complex signals, there is an increasing trend to hybridize the architectures of well-performed deep learning algorithms to leverage their strengths. For the RUL prediction of cutting tools, signals could be from different sensor sources demonstrating various physical features. It is worth exploring how to design an appropriate hybrid deep learning algorithm to fuse signals and their inherent features effectively in supporting more accurate RUL estimation of cutting tools.

Meanwhile, the complexities of signals for a cutting tool arise not only from different types of sensors but also from different stages of a cutting tool lifecycle. Features in the stages could be in dramatically unbalanced distributions. To further improve the performance of the deep learning algorithm in terms of accuracy and computational efficiency, it is imperative to develop an effective strategy to segment a large volume of signals wisely to facilitate the deep learning algorithm for signal processing.

This chapter is aimed at addressing the above aspects. The system proposed in this research consists of two subsystems: (i) to enhance the prediction accuracy and expedite computational efficiency, a Hurst exponent-based method is developed to partition complex signals along tool wear evolution; (ii) a hybrid CNN-LSTM algorithm, which combines feature extraction, fusion and regression, is designed and applied for predicting the RUL of cutting tools. The system is validated using a case study with a large set of signals from multiple sources. The innovative characteristics of the research are below:

A novel systemic methodology is designed to integrate strategies of signal partition and deep learning for effectively processing and analyzing multi-sourced sensor signals throughout the lifecycle of a cutting tool;

Detailed comparisons between the proposed system and other main-stream algorithms are conducted, and benchmarks are made to demonstrate the superior performance of the research.

The rest of this chapter is organized as follows. The related works are reviewed in Sect. 2. The developed methodology is presented in Sect. 3. Section 4 presents methodology validation and benchmarking with comparative algorithms. Finally, conclusions are drawn in Sect. 5.

2 Literature Survey

Flank wear, which occurs on the surface of a cutting tool that contacts with a workpiece, is the most common form of tool wear [7]. A lot of related works conducted analysis on flank wear as it is able to provide a more reliable RUL prediction on a cutting tool. To simplify the terminology, the flank wear of a cutting tool is called tool wear in the rest of the chapter. As aforementioned, there are various dynamic and uncertain factors in practical machining processes, and an experiment-based approach is not effective in evaluating tool wear and estimating RUL effectively. Instead, a data-driven approach has been receiving more attentions considering the approach is more robust in supporting practical applications. The performance of the approach is highly dependent on the embedded intelligent algorithms and the mechanisms for signal processing.

Zhang et al. [7] collected vibration signals in a machining process, and key features were identified based on Pearson correlation coefficient (PCC). A neural-fuzzy network (NFN) was developed to predict tool wear and the corresponding RUL. The research provided comparisons between the NFN and an ANN, and between NFN and a radial basis function network (RBFN). Based on the results, it was concluded that the NFN had a better performance on the prediction of tool wear and RUL. Yu et al. [8] developed a hidden Markov model (HMM)-based method for tool wear monitoring and RUL prediction. The root mean square (RMS) of vibration signals was calculated as a health indicator to construct a multiple HMM in a weighted manner for RUL prediction. In the method, tool wear was divided into discrete regions by the HMM to better envision the status and evolution of tool wear. Wu et al. [9] proposed a multi-sensor fusion system for online prediction of RUL for cutting tools. In the method, first, the statistical features in the signals of vibration, acoustic emission and current in the time- and frequency- domains were extracted. In considering the relationship between time-varying signals and cutting tool wear, features were manually segmented, and correlation analysis, monotonicity analysis and residual analysis were carried out for selecting the most relevant features to support the prediction. Then, an adaptive network-based fuzzy inference system (ANFIS) was applied to achieve feature fusion. Finally, the RUL prediction was conducted using a polynomial curve fitting algorithm. Yang et al. [10] established a model for predicting tool RUL based on a trajectory similarity-based prediction (TSBP) algorithm and the differential evolution SVR (DE-SVR) algorithm. Signals of cutting force were collected during a machining process, and analyzed in the time domain, frequency domain and wavelet to extract the most representative features. The root mean square value (RMSV), average value (AV) and the standard deviation value (SDV) in the time- and frequency- domains were evaluated to investigate their correlation with tool wear. Then these features were identified via its importance to be further imported into an integrated model for prediction.

In recent years, deep learning algorithms were leveraged to support more effective analysis and reasoning for the RUL prediction on cutting tools. For example, An et al. (2020) proposed an integrated model that combined a CNN and a stacked bi-directional and unidirectional LSTM (SBULSTM) to predict the RUL of a cutting tool. Among them, CNN was used to extract features from signals for dimensionality reduction, and then SBULSTM was used to train these features and achieve the purpose of prediction. More works can refer to recent reviews on the topics [2, 3].

Although the reviewed works achieved some success in the tool RUL prediction, there are still obvious defects and potential improvement areas below:

To meet more complex industrial requirements and enhance the prediction accuracy, multiple types of sensors are increasingly adopted and a large amount of signals will be generated during processing. The approach based on a single machine learning algorithm could be hindered in processing high-dimensional inputs [11]. With the growth of data, algorithms built could be no longer a single network architecture for data analysis. Recently, the study of more complex hybrid network models has been becoming a new trend [12]. Such the state-of-the-art strategy integrates the strength of various learning algorithms and presents greater advantages to process the data sources that contain multiple characteristics. Among the hybrid models, a hybrid CNN-LSTM algorithm is promising in performance and was applied in some applications, such as action video recognition [13], traffic forecasting [14], residential energy consumption [15], engineering [16]. Based on the evidences of successful applications, it is valuable to consider designing an appropriate CNN-LSTM algorithm to perform the RUL prediction of cutting tools with a high performance.

Moreover, due to the imbalanced distributions of acquired signals during a processing, deep learning algorithms could lead to inaccurate prediction throughout the tool lifecycle. Therefore, it is useful to develop sensible strategies to segment sensor signals into fragments to promote the efficiency and reliability of deep learning algorithms in processing the signals. Low et al. [17] showed that the multi-model prediction of the RUL of a cutting tool based on several discrete partitions of data had a better performance than working on the entire data directly. Similarly, Opałka et al. [18] demonstrated that a more accurate RUL prediction can be achieved by dividing time series sensor data into smaller segments according to a certain indicator. The Hurst exponent, which was developed by Hurst (1900–1978) based on the rescaled range (R/S) analysis method, has been extensively applied to determine the long-term memory of time series data that are varying in degree and time span [19]. There are increasing uses of the Hurst exponents in signal segmentation to support different applications [20, 21]. In this research, an effective strategy of the Hurst exponents is explored in establishing correlations between signals and tool wear development, then a sensible partitioning method for the signals is developed.

3 Methodology and System Flow

The workflow of the system and the methodology for the RUL prediction on cutting tools are shown in Fig. 1.

The system consists of two subsystems, i.e., the Hurst exponent-based data partition, and a hybrid CNN-LSTM algorithm for RUL prediction. In the system, three types of sensors are deployed to monitor the statuses of a cutting tool, i.e., vibration (V), cutting force (F) and acoustic emission (AE). The signal partition method based on the Hurst exponent is employed to segment sensor data into small batches to minimize the impact of feature imbalances from the input data. The partitioned datasets are then sent to the CNN-LSTM algorithm for processing, where the RUL prediction on a cutting tool is generated. In the CNN-LSTM algorithm, each individual CNN is designed as a channel to process one type of signals, and the extracted features from the signals are fused in a concatenated layer, and the synchronized signals are further sent to the LSTM as a regression layer for predicting tool wear and RUL.

3.1 Signal Partition Based on the Hurst Exponent

The input signals for the developed system are from multiple sensors. Multiple sensors V, F, and AE are mounted on a CNC machine. After a total k cuts are executed, each sensor collected N samples of data in each cut. The raw datasets $\{M^i | i = 1, 2, \cdots k\}$ are organized to obtain the sub-datasets for each sensor. Then, the Hurst exponent is conducted for partition on the data of individual cut from each sensor to acquire the input for the CNN-LSTM algorithm.

Signals from different sensors could have different impacts on the RUL prediction. The Hurst exponent is employed to process various sensor signals in order to establish accurate correlations between the signals and the prediction of tool wear.

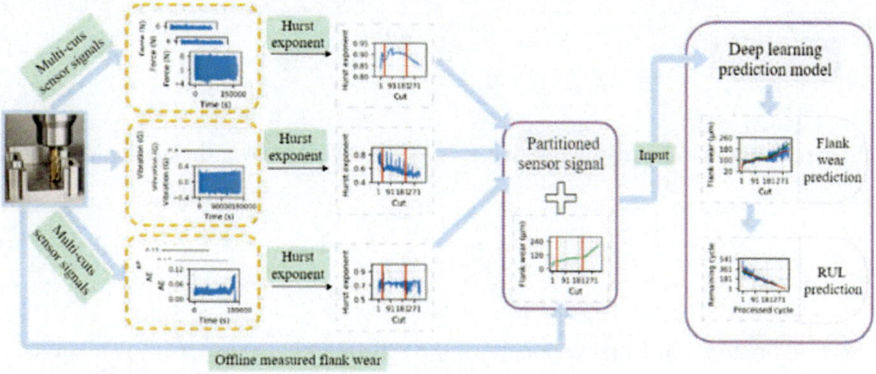

Fig. 1 The flow and architecture of the presented system

Key parameters of the Hurst exponent for sensor signals can be obtained using the following steps:

For each window size n between N, N/2, N/4, N/8…until n approaching 350, repeat Step 1 – Step 6:

Step 1: Divide a given time series of the sensor signal for each cut into M subseries of length n. Then calculate the mean value $(\bar{T}_m)$ of the mth subseries as follows:

$$\bar{T}_m = \frac{1}{n}\sum_{i=1}^{n} T_{i,m} \tag{1}$$

where, $T_{i,m}$ stands for the ith signal in the mth subseries and $m = 1,2\ldots M$.
Step 2: Create mean adjusted series D_m of the sensor signal $T_{i,m}$:

$$D_{i,m} = T_{i,m} - \bar{T}_m fori = 1, 2\ldots n \tag{2}$$

Step 3: Calculate the cumulative deviate series Z_m:

$$Z_{j,m} = \sum_{j=1}^{i} D_{i,m} fori = 1, 2\ldots n \tag{3}$$

Step 4: Calculate the range of R_m:

$$R_m = \max(Z_{1,m}, Z_{2,m}, \ldots, Z_{n,m}) - \min(Z_{1,m}, Z_{2,m}, \ldots, Z_{n,m}) \tag{4}$$

Step 5: Calculate the standard deviation, S_m, by using the following:

$$S_m = \sqrt{\frac{1}{n}\sum_{i=1}^{n} (T_{i,m} - \bar{T}_m)^2} \tag{5}$$

Step 6: Calculate the rescaled range R_m/S_m and average over the all the M samples:

$$R/S = \frac{1}{M}\sum_{m=1}^{M} R_m/S_m \tag{6}$$

Step 7: Finally, the Hurst exponent of the sensor signal of each cut is obtained by linear fitting the log values of the rescaled ranges R/S for all the window sizes n based on:

$$\log(\frac{R}{S}) = \log C + H \cdot \log n \quad (7)$$

where C is a constant, H is the Hurst exponent.

The value of a Hurst exponent (i.e., H value) varies between 0 and 1, and it indicates the dependence of sensor signals on their past values [22]. Based on the previous research, the following observations are made for various signals (e.g., V. F and AE in this research):

(a) When $0 \leq H < 0.5$, there is a negative correlation between the sensor signals and flank wear. A smaller H value implies that the signal fluctuation in this time period changes more dramatically.
(b) When $H = 0.5$, the corresponding signal presents the Brownian motion. It refers that the signal has no impact on future signals, and signal fluctuation will be completely random and unpredictable.
(c) When $0.5 < H < 1$, the fluctuation of the signal shows a continuous positive correlation between the sensor signals and flank wear. The future trend will follow the changes of the present signal, and this type of serial signal is predictable.
(d) When $H = 1$, time series signals will be in a straight line and there is no fluctuation and correlation.

Within the lifecycle of a cutting tool, the value of a Hurst exponent is obtained from each signal by using the above steps. The different values of the Hurst exponents will be used to establish the correlation between signals and various flank wear stages of a tool lifespan, which are normally divided into an initial region of rapid wear, a steady region of uniform wear and a severe region of dramatically increased wear. Accordingly, the sensor signals can be segmented based on the stages to facilitate the following prediction using a deep learning algorithm.

3.2 A Hybrid CNN-LSTM Algorithm for Prediction

CNN has demonstrated good performance at intelligent acquiring features from data and is immune to the frequency variation in the data. LSTM is more powerful for processing time series data. However, both the algorithms present their own limitations when dealing with real-time data. CNN assumes that all inputs and outputs are independent of each other in processing. The inherent information between features are neglected, resulting in performance degradation when processing time series data [23]. LSTM can extract long-term dependencies of features in the data sequence to improve recognition accuracy. However, the training time for LSTM is much longer than that of CNN as it needs to learn the nonlinearity relationships in the data. To leverage the characteristics of both algorithms, in this work, a hybrid CNN-LSTM algorithm is designed for better prediction on tool life. It is capable to identify the spatial and temporal relations in sensor signals. Based on the different types of signals, multiple CNN sub-models are constructed as pre-processors. The CNN

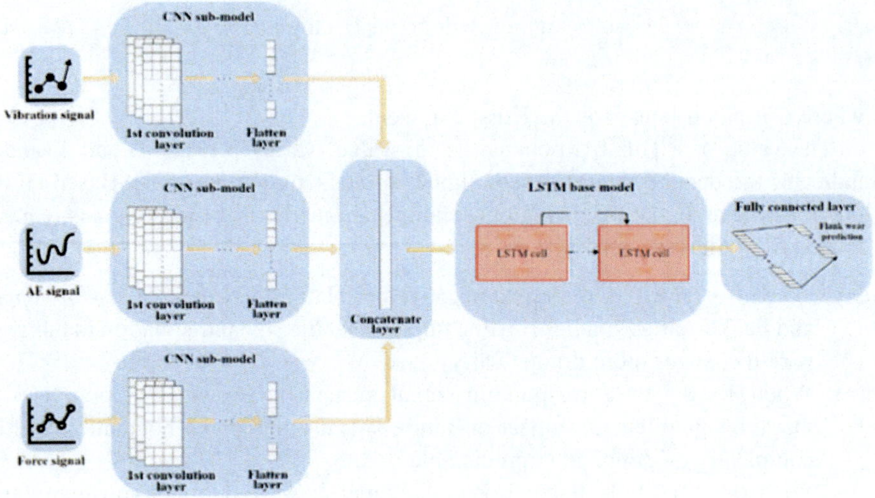

Fig. 2 The flow and architecture of the presented system

sub-model parallelly extracts the features from each sensor node, and these features are then fused at a concatenate layer. Finally, the concatenated features from sensor signals are sent to the LSTM for prediction. The proposed CNN-LSTM model is illustrated in Fig. 2.

CNN design.

CNN, which is one of the most successful deep learning models, offers the strength of extracting high-level dependency features automatically from input data. The learning ability and the training time of the CNN model are decided by its structure, especially the number of layers. Normally, a shallow structure cannot provide sufficient processing performance, and meanwhile an excessive deep CNN may be harmful to the time sequential aspect of the data or cause overfitting.

According to the different dimensionality of the input variables, CNN is commonly classified into 1D and 2D structures, which the 1D CNN is mainly used for sequential data processing, and 2D CNN is more suitable for image recognition [24]. In the field of machinery condition prognosis, these two CNN architectures have been employed on the basis of the sensor signal. For 2D CNN, the format conversion of the time series data is necessary for matching the use of 2D CNN, which have been performed in the work of [25]. Although, the conversion brings positive effects on the accuracy improvement of the prediction, the technology is complex and requires prior knowledge, and it is perhaps the advised option only if there is an additional need [23]. On the contrary, to process the machinery time series signal with 1D CNN has been proved more appropriate in several related studies [24, 25]. 1D CNN better utilizes its characteristic of automatic feature extraction and avoids the deviation caused by manual signal processing. Moreover, based on the prediction model architecture established in this work, the subsequent regression prediction of

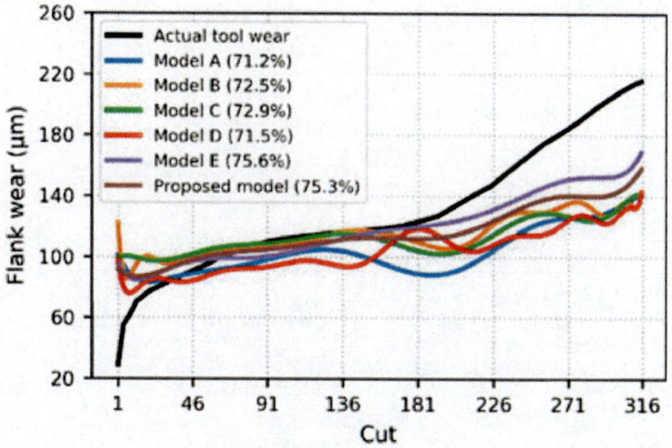

Fig. 3 The prediction accuracies of the CNN models

LSTM seeks the time series data as the input, rather than other formats, to preserve the temporal integrity of the sensor signal.

Thanks to the potential of all aspects of 1D CNN in processing sensor signals, a 1D CNN structure is adopted in this research. In order to achieve a satisfactory prediction accuracy and efficiency in processing the sensor signals, the 1D CNN model with different numbers of convolution and pooling layers are evaluated to identify the most suitable architecture. Based on the same input dataset (the exact dataset is described in Sect. 4), the prediction accuracies of each 1D CNN model are shown in Fig. 3.

Based on the above results, it clearly shows that, within the six 1D CNN architectures, the best prediction accuracy was achieved by the model E, which is 75.6%. The accuracy of the proposed model of three convolutional layers and two pooling layers is 75.3%, which is slightly lower than that of the model E. However, the model E with the six-layers structure requires much longer computing time. Thus, the proposed model provides the best trade-off in accuracy and efficiency performance among the six models in this research. The exact configurations of the proposed model are shown in Fig. 4. To process three types of sensor signals, i.e., V, F and AE, there are three such CNN models arranged in parallel (i.e., the CNN models of 1, 2 and 3 for the signals of V, F and AE respectively).

Each partitioned sensor signal in a format of matrix is fed into the CNN model through its input layer. For segmented sub-matrix for the inputs of the CNN models of 1, 2 and 3, they are below:

$$A = \begin{bmatrix} \mathrm{a}_1^1 \ \mathrm{a}_2^1 \ \mathrm{a}_3^1 & \cdots \mathrm{a}_n^1 \\ \mathrm{a}_1^2 \ \mathrm{a}_2^2 \ \mathrm{a}_3^2 & \cdots \mathrm{a}_n^2 \\ & \vdots \\ \mathrm{a}_1^d \ \mathrm{a}_2^d \ \mathrm{a}_2^d & \cdots \mathrm{a}_n^d \end{bmatrix} B = \begin{bmatrix} \mathrm{b}_1^1 \ \mathrm{b}_2^1 \ \mathrm{b}_3^1 & \cdots \mathrm{b}_m^1 \\ \mathrm{b}_1^2 \ \mathrm{b}_2^2 \ \mathrm{b}_3^2 & \cdots \mathrm{b}_m^2 \\ & \vdots \\ \mathrm{b}_1^d \ \mathrm{b}_2^d \ \mathrm{b}_2^d & \cdots \mathrm{b}_m^d \end{bmatrix}$$

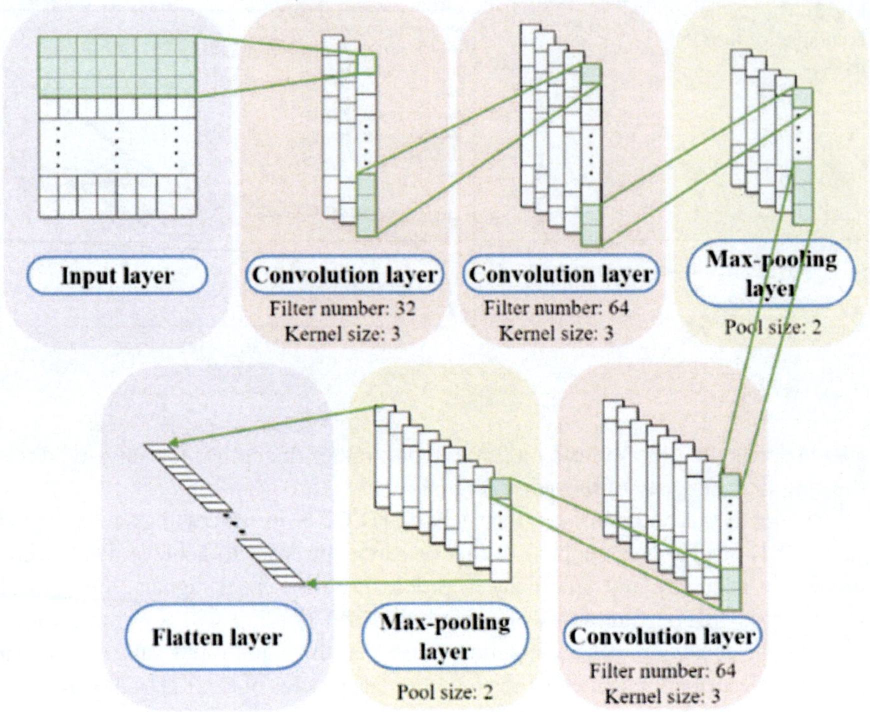

Fig. 4 The proposed CNN architecture

$$C = \begin{bmatrix} c_1^1 & c_2^1 & c_3^1 & \cdots & c_e^1 \\ c_1^2 & c_2^2 & c_3^2 & \cdots & c_e^2 \\ & & & \vdots & \\ c_1^d & c_2^d & c_2^d & \cdots & c_e^d \end{bmatrix} \tag{8}$$

where, *n*, *m*, *e* denotes the number of signals of sensor *A*, *B* and *C*, respectively, *d* denotes the number of cuts in the dataset after segmentation.

The convolution layer of each CNN model convolutes the input matrix to generate the spatial feature map by the activation function, and it can be described as:

$$\Phi_l = f(\mathrm{conv}(X_{l-1} * \mathrm{w}_l) + \mathrm{b}_l)) \tag{9}$$

where, the Φ_l denotes the feature map of the lth convolution layer, X_{l-1} denotes the generated feature map from $(l-1)$ th layer, w_l denotes the weight, b_l denoted the bias, the conv(∗) denotes the convolution process. The $f(\cdot)$ denotes the activation function. The rectified linear unit (ReLU) is selected as the activation function for each convolution layer in the proposed CNN model. The reason is that the ReLU increases the nonlinearity between layers, and only a small amount of computation is needed to alleviate the gradient vanishing problem (Carneiro et al. 2016).

After the convolution layer, the pooling layer is applied to further reduce the feature dimensionality, the max-pooling function is selected for every pooling layer, which extracts the maximum feature value with a window of size 2, so as to retain the important feature information and improve training efficiency. The output feature map can be depicted as:

$$\Phi_k = \mathrm{w}_k \cdot \max(\Phi_l) + \mathrm{b}_k \tag{10}$$

where, the Φ_k denotes the feature map of kth pooling layer, w_k denotes the weight, b_k is the bias, $\max(\cdot)$ denotes the max-pooling function.

Finally, the flatten layer is connected to the last pooling layer, to complete the transition from the convolution layer to the next layer base model by converting the generated features into a one-dimensional array. And the size of the output array equals the number of the cuts d.

So far, the signal features extracted by the CNN model eliminate the interference between the heterogeneous sensors, and these features can be fused without complicated operations. Therefore, the concatenation layer is applied to be responsible for the feature fusion after the multi-channel CNN models. As the element number of each feature vector is d, the feature fusion can be depicted as:

$$\begin{bmatrix} \mathrm{a}'_1 \\ \mathrm{a}'_2 \\ \vdots \\ \mathrm{a}'_d \end{bmatrix} + \begin{bmatrix} \mathrm{b}'_1 \\ \mathrm{b}'_2 \\ \vdots \\ \mathrm{b}'_d \end{bmatrix} + \begin{bmatrix} \mathrm{c}'_1 \\ \mathrm{c}'_2 \\ \vdots \\ \mathrm{c}'_d \end{bmatrix} \rightarrow \begin{bmatrix} \mathrm{a}'_1 & \mathrm{b}'_1 & \mathrm{c}'_1 \\ \mathrm{a}'_2 & \mathrm{b}'_2 & \mathrm{c}'_2 \\ \vdots & \vdots & \vdots \\ \mathrm{a}'_d & \mathrm{b}'_d & \mathrm{c}'_d \end{bmatrix} \tag{11}$$

where, $a^{'}$, $b^{'}$ and $c^{'}$ denotes the feature of different sensor generated from the CNN models.

LSTM

The focus of LSTM is to capture the dependencies of extracted features to achieve prediction. Once the data features of each CNN model have been merged at the concatenated layer, the generated column-wise array will be the input of the LSTM model, which can be expressed as $X = \left[x_{i,j}\right]$, where x denotes the feature value,$i = \{1, 2, \cdots d\}$, j denotes different signals. A multi-layer LSTM uses X as input data, and each layer contains a LSTM cell (Fig. 5).

For the current time-step t, the output h_t, which indicates the predicted flank wear value, and the memory state c_t of LSTM cell are decided by the new input x_t, the output h_{t-1} and memory state c_{t-1} of the last time step $t-1$. Based on the input array X, which contains d time-steps and j features in each time-step, the LSTM network will recursively obtain the comprehensive information from each time-step and provide corresponding predictions.

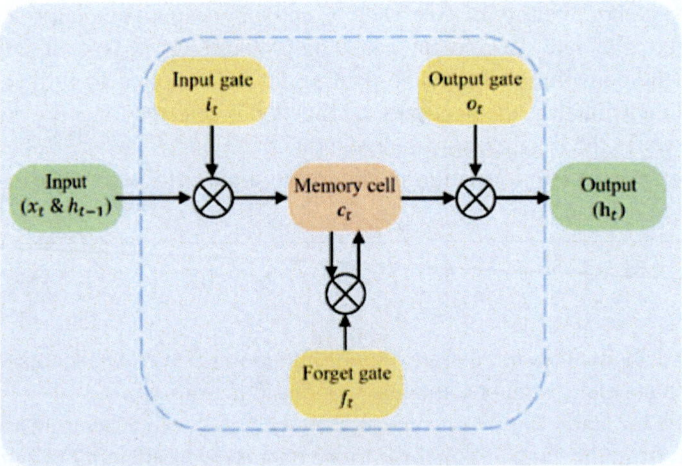

Fig. 5 The proposed CNN architecture

During the process, the forget gate f_t of the LSTM cell decides how many information of c_{t-1} should be forgotten according to the sigmoid activation function, and thus create the new feature data as candidate value. Then the candidate value is fed into the input gate i_t to update the memory cell state. Finally, the prediction value h_t of the LSTM cell at the time-step t can be calculated based on the updated cell state, which is from c_t and controlled by the output gate o_t. The above gate and cell can be obtained as follows:

$$i_t = \sigma(W_i x_t + U_i h_{t-1} + b_i) \tag{12}$$

$$i_t = \sigma(W_i x_t + U_i h_{t-1} + b_i) \tag{13}$$

$$o_t = \sigma(W_o x_t + U_o h_{t-1} + b_o) \tag{14}$$

$$c_t = f_t \odot c_{t-1} + i_t \odot \tanh(W_c x_t + U_c h_{t-1} + b_c) \tag{15}$$

$$h_t = o_t \odot \tanh c_t \tag{16}$$

where, σ is the sigmoid activation function; $\odot$ is the Hadamard product; W and U are variable weights and b is the bias.

Multiple LSTM cells are connected to form an LSTM network in time order, to predict the time sequence output (flank wear), that is $\{h_1, h_2, \cdots h_d\}$. The number of the LSTM cell depends on the number of the feature in each signal received from the concatenation layer. The architecture of the LSTM model is shown in Fig. 6.

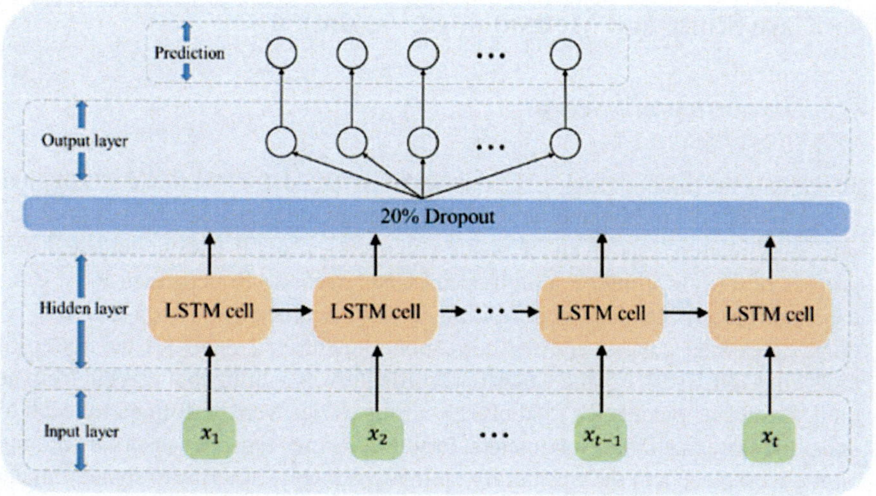

Fig. 6 The LSTM architecture in this research

Since the input data is partitioned via Hurst exponent into sub-datasets in this chapter, the model training based on size-reduced sub-datasets may increase the possibility of overfitting. Therefore, inhibiting the overfitting of the proposed hybrid model is necessary to further improve the prediction performance. As the dropout is one of the most popular and efficient regularization technologies used to prevent the overfitting, it has been employed in the LSTM model. Dropout randomly discards the hidden neuron with the setting dropout rate, and the remaining neuron is trained via backpropagation to obtain the new weight and bias. It can be described as:

$$S_i = \sum \sum_j w_{ij} p_j S_j \tag{17}$$

where, the S_i denotes the output of ith layer after the dropout, S_j denotes the output of the j th layer (previous layer), w denotes the weight, p denotes the dropout rate, which is set to 0.5 in this chapter to maximize the regularization.

Furthermore, in the last output layer of the hybrid model, the linear activation function is adopted to implement the regression prediction. For the prediction evaluation of the proposed hybrid CNN-LSTM model, the mean absolute error (MAE) is used as the criteria, which quantifies the absolute error between the prediction and actual values. It can be expressed as:

$$MAE = \frac{1}{n} \sum_{i=1}^{n} \left| \tilde{y} - y \right| \tag{18}$$

where, n is the training sample size; $\tilde{y}$ is the prediction value; y is the actual value.

4 Case Study and Methodology Validation

4.1 Experimental Setup

To validate the methodologies, the dataset from the 2010 PHM Society Conference Data Challenge was adopted here. The dataset includes signals from cutting force sensors, vibration sensors and acoustic emission sensors that were collected in the process of a dry milling on a high-speed CNC machine Roders Tech RFM760. In the experiment, a stainless-steel workpiece was machined using six 3-flute ball nose tungsten carbide cutters, and the corresponding sensor signals for the six cutting tools C1, C2, C3, C4, C5 and C6 were recorded. Every cutting tool was cut from new until significant wear, and a total of 315 cutting cycles were performed for each tool under the same machining parameters. In addition, three types of sensors are mounted on the workpiece and the machining table respectively, in terms of dynamometers, accelerometers and an acoustic emission sensor. The schematic of the experiment is shown in Fig. 7.

During the machining process, three Kistler quartz 3-component platform dynamometers, three Kistler piezo accelerometers and a Kistler acoustic emission (AE) were used. These seven signal channels are shown in Table 1. A cutting tool processed the workpiece surface line-by-line along the x-axis with the axial depth of 0.2 mm, radial depth of 0.125 mm and cutting length of 108 mm per cut until the entire surface was removed. The spindle speed was maintained at 10,400 RPM and

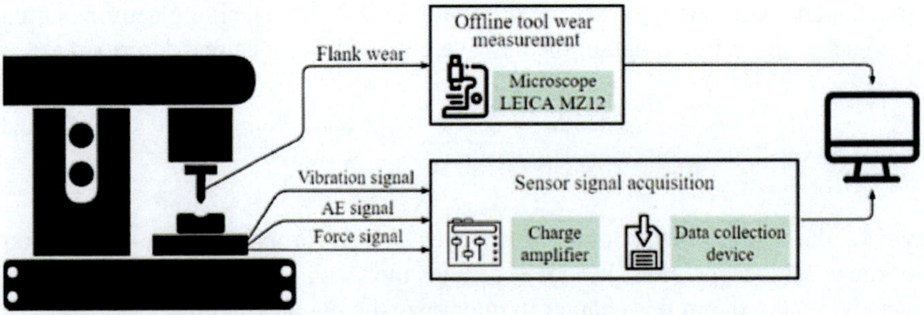

Fig. 7 The LSTM architecture in this research

Table 1 Sensor signal of the dataset

Sensor type	Signal sources
Kistler quartz dynamometer	Force (N) in the X axis, Force (N) in the Y axis, Force (N) in the Z axis
Kistler piezo accelerometers	Vibration (g) in the X axis, Vibration (g) in the Y axis, Vibration (g) in the Z axis
Kistler acoustic emission	AE-RMS (V)

the feed rate was 1,555 mm/min. Moreover, after each cut, the cutting tools C1, C4 and C6 were placed under a LEICA MZ12 microscope to measure the flank wear (Vb) of each flute, and three datasets that combined with sensor signals and flank wear of cutting tool C1, C4 and C6 were employed in this work. The dataset size of a single cutting tool is approximately 3.2 GB.

4.2 Signal Partition Based on the Hurst Exponent

Based on the datasets of the acquired sensor signal, the validation of the proposed system is implemented on a 3.60 GHz Intel (R) Core (TM) i7-7700 CPU processor (with 8.00 GB of RAM), in which, the data processing and the computation of the deep learning algorithm are executed on the Keras framework and Tensorflow, respectively.

As aforementioned, 315 cuts were performed using the cutting tools C1, C4 and C6, respectively. Taking C4 as an example for analysis, the dynamometer signals on the X axis, vibration signal on the X axis and the AE sensor signals for the 1st cut and the last (315th) cut are plotted in Fig. 8. It can be observed that the amplitude of the three sensor signals increases along with the machining process, indicating

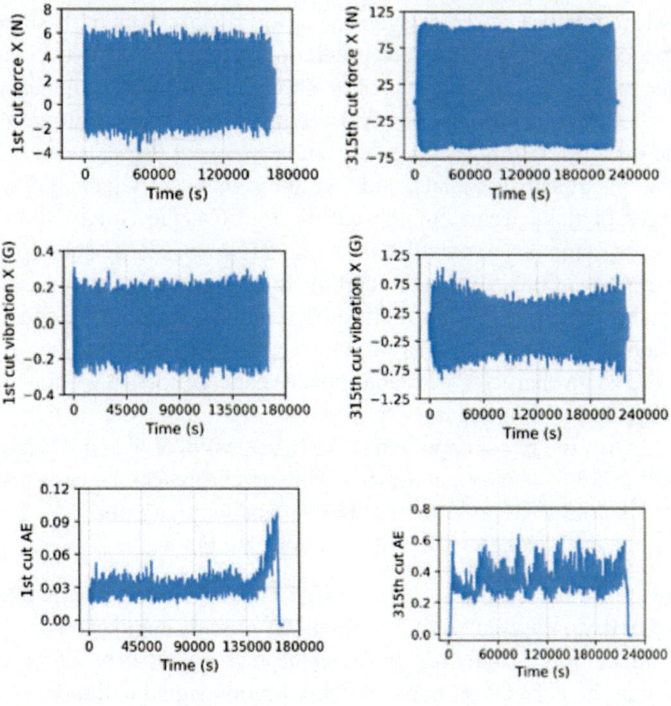

Fig. 8 Sensor signals for the 1st and 315th cut by the cutting tool C4

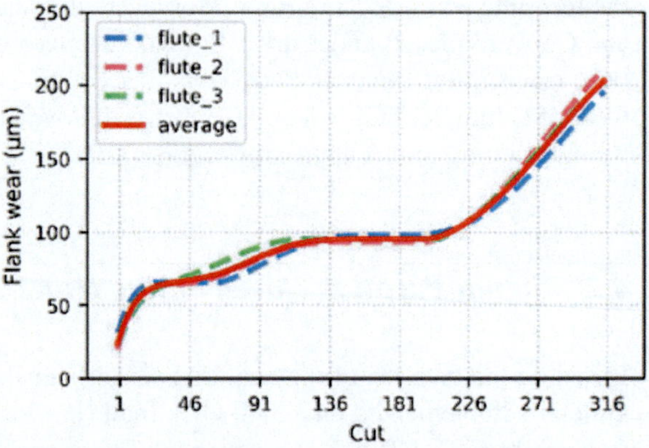

Fig. 9 The flank wear of the cutting tool C4

that the signals exhibit a good correlation with tool wear deterioration. Thus, these sensor signals are feasible to establish the prediction.

Corresponding to each cut, the flank wear of three flutes on the cutting tool was measured in the experiment. The flank wears of the cutting tool C4 are shown in Fig. 9. According to the recommendation of the ISO 8688–2 (1989), the cutting tool life criterion is commonly predetermined by the average wear value of all flutes. Therefore, the average flank wears of the cutting tools C1, C4 and C6 were used for training the prediction algorithm presented in this chapter.

Tool wear involves different stages. Sensor signals over each stage is usually uneven, which may cause the wear prediction inaccurate. Moreover, as discussed earlier, due to the changing features of tool wear in the different stages, it is difficult for the prediction algorithm to effectively estimate different wear trends of tool wear during these stages. In this research, the Hurst exponent is used as an index to judge the fluctuations of sensor signals, and then the signals are segmented to correspond to the stages of flank wear. For the cutting tool C4, the Hurst exponent of three types of sensor signals are calculated for the 315 cuts. The Hurst exponents of the vibration signals in the X-axis, the cutting force signals in the X-axis and the AE signal are shown in Fig. 10. Furthermore, to better represent the convergence and visualization effect of the results, a cubic curve, which is a regression analysis method that preserves data characteristics and reduces data turbulence without changing the data, was applied to fit the Hurst exponent.

From Fig. 10, the Hurst exponents (the H values) of V, F and AE are all roughly between 0.5 and 1. It means that these sensor signals present the persistent behaviors along with the tool wear and it is feasible to conduct prediction.

For the vibration signals, some observations are below:

- The H value of the cutting tool is the biggest at the beginning of machining, which is close to 0.85, implying that the signal has obvious regularity;
- As the machining progresses, the H values are higher than 0.6 before the 20th cut, so that the correlation between the vibration signal and the tool wear is still strong at this time. From the 20th cut to the 205th cut, the H values decrease to

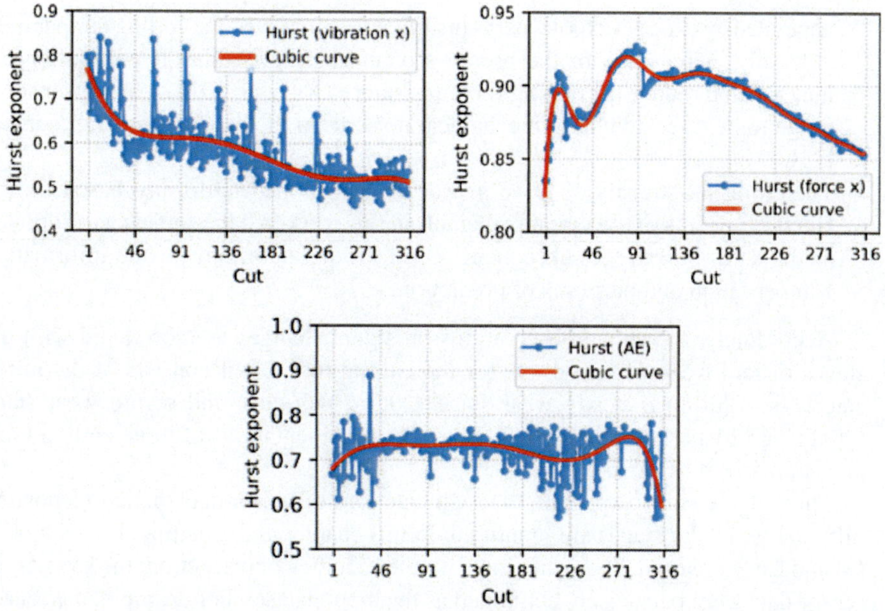

Fig. 10 The Hurst exponent of the sensor signals (V, F and AE) for the cutting tool C4

between 0.5 and 0.6, which indicates that the long-term memory of the signal is lower than its previous tool wear stage;

- After the 205th cut, the H values progressively approach to or even are lower than 0.5, meaning that the signals exhibit a trend of the Brownian motion. That is, the probability of negative correlation between the signal and the tool wear is increased;
- After the 315th cut, the H values are bound to drop below 0.5, so that the signal will completely show a negative correlation, and the cutting tool exceeds its health lifespan and the tool wear displays unpredictability.
- Overall, the fractal of the H values trend of the vibration signal has a strong correspondence with the different stages of tool wear, and the change of the H values are more sensitive. Thus, the vibration signal displays the greatest potential among the three signals for signal partition.
- For the signals of the cutting force and AE, some observations are below:
- Along with the machining process until the 315th cut, the H values both show a decreased trend, which corresponds to the tool wear. Moreover, the H values are both greater than 0.5, it represents these sensor signals exhibit persistent behaviors and are positively correlated with the tool wear, thereby revealing that prediction of tool wear based on the cutting force and AE signals are also feasible;
- Despite this, it should be noted, the Hurst exponent trend of these signals are not as clear as the vibration signal in performing partition on the signals;
- This may be interpreted as: due to the acquisition frequency of the cutting force and AE signal is lower than that of the vibration signal, the collected noise signal that

generated by excessive tool wear is insufficient. Moreover, the cutting temperature gradually increases with the progress of machining, so that it will soften the material and reduce the cutting force resistance (Xu et al. 2018). An AE sensor is prone to be affected by the mechanical noise from the background environment [26];

- Therefore, the signals of three sensor are partitioned uniformly based on the Hurst exponents of the vibration signals in this work. All the signals were applied as the data sources for subsequent deep learning algorithm for the information compensation and purposes of prediction.

Accordingly, the above observations on signal changes are correlated with the stages of tool wear. For instance, for the cutting tool C4, it consists of the initial stage (1st–19th cut), steady wear stage (20th–204th cut), and severe wear stage (205th–315th) as shown in Fig. 11. The signal segmentations of three cutting tools are summarized in Table 2.

Through the Hurst exponent, the segmentations of sensor data that correspond to different tool wear stages are identified. In this chapter, three cutting tools, i.e., C1, C4 and C6, were used. To further improve the accuracy of prediction, the segmented sensor data were pair-wisely combined as the input dataset. For example, the sensor signal and flank wear of C1 and C4 were combined as a training dataset (denoted as C1C4), and the sensor signal and flank wear of C6 was treated as the validation dataset at the same time (refer to Table 3).

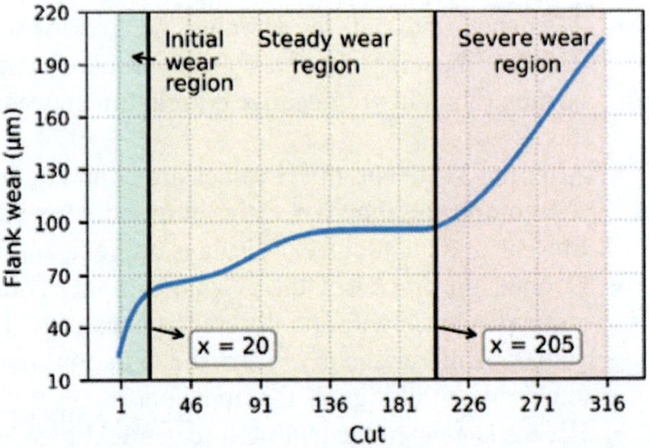

Fig. 11 The stages of flank wear for the cutting tool C4

Table 2 Signal segmentation of each cutting tool

Cutting tool	Tool wear stages		
	Initial wear region	Steady wear region	Severe wear region
C1	1 to 49	50 to 139	140 to 315
C4	1 to 19	20 to 204	205 to 315
C6	1 to 14	15 to 179	180 to 315

Table 3 Comparisons of prediction accuracy for partitioned and un-partitioned datasets

Dataset	Test	Prediction accuracy				
		Partitioned dataset				Un-partitioned dataset (%)
		Initial stage (%)	Steady stage (%)	Severe stage (%)	Integrated (%)	
C1C4	C6	90.7	89.8	83.8	88.1	77.9
C1C6	C4	87.4	88.6	87	86.0	75.6
C4C6	C1	89	87.8	86.3	87.7	73.2

4.3 Performance Evaluation on the Hurst Exponent and CNN-LSTM

To evaluate the performance of the Hurst exponent-based partition, un-segmented signals (raw signals of each sensor) were used to perform prediction on flank wear based on the designed CNN-LSTM algorithm. The prediction accuracy of each dataset in this work is obtained by its corresponding validation set. And the prediction curves for the sensor signals based on the CNN-LSTM algorithm are shown in Fig. 12.

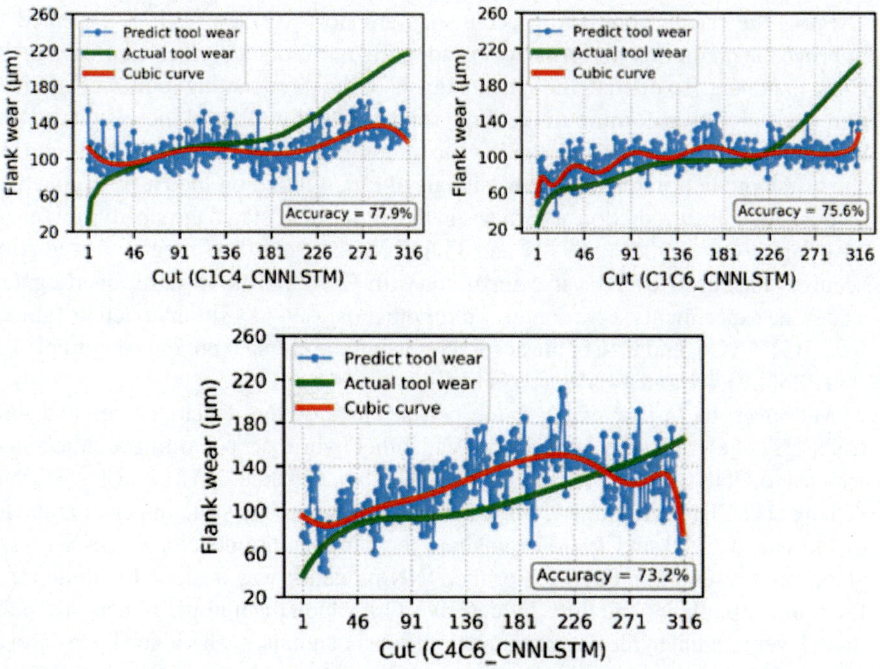

Fig. 12 Prediction curves of the sensor signals based on the CNN-LSTM

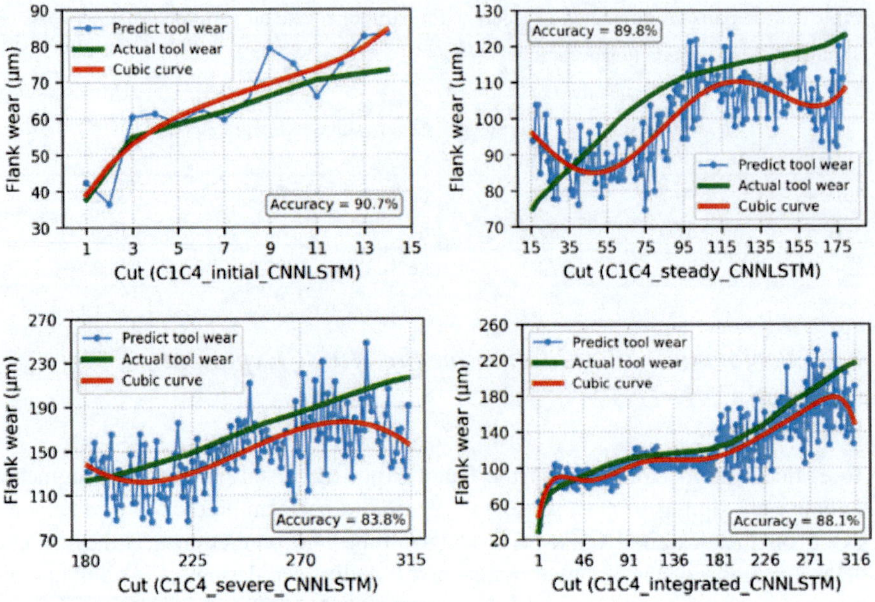

Fig. 13 Prediction results based on the partitioned signals for C1C4

From the results, prediction accuracies are not satisfactory. It is expected to improve the prediction performance by adopting partitioned dataset according to the Hurst exponent. To do that, the sensor signals of the three cutting tools were divided into three groups according to the three stages of tool wear, and these signals were combined with corresponding flank wear values to form the new datasets. Figure 13 shows the prediction results for the signals of C1C4 based on the Hurst exponent.

For C1C4, the prediction accuracies in the initial wear stage, the steady and severe wear regions were 90.7%, 89.8% and 83.8%, respectively. The integrated prediction accuracy increased to 88.1% in comparison with 77.9% for the un-partitioned signals. The same experiments were conducted for other datasets. As summarized in Table 3, for C1C4, C1C6, and C4C6, the accuracies using the Hurst exponent were improved by 10.1%, 10.4% and 14.5%, respectively.

Moreover, to further evaluate the performance of the developed research, the CNN-LSTM algorithm was compared with other hybrid deep learning architectures, which are CNN-CNN [27], LSTM-LSTM (Choi and Lee 2018) and DNN-DNN (Zhang et al. 2019). To better compare the architectures, the structure and parameter of individual CNN and LSTM remained the same as the developed CNN-LSTM algorithm presented in Sect. 3. For the DNN model, it was designed with an input layer, an output layer and three hidden layers, in which the number of neurons in the input layer is equal to the sample number of sensor signals, each hidden layer is set as twice the number of input signal after multiple tests, and the size of the output layer depends on the number of flank wear values. The structure of the DNN is shown in Fig. 14.

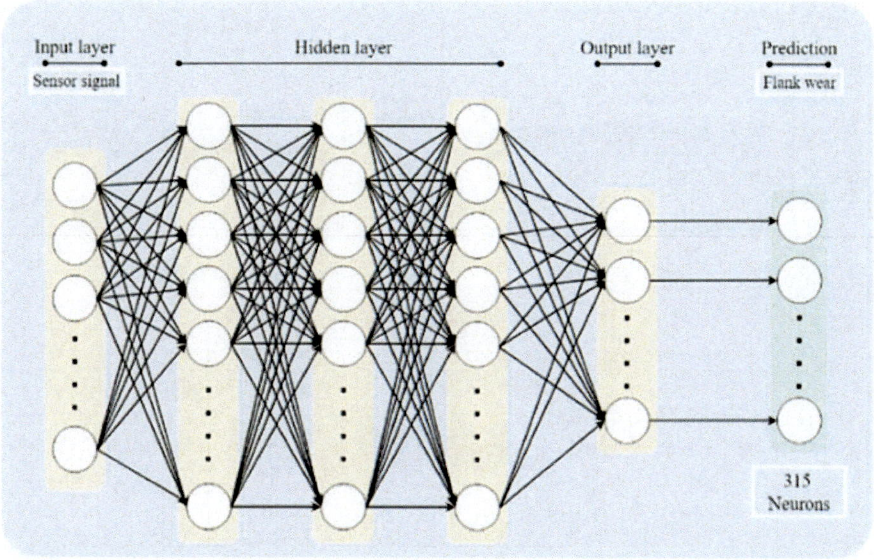

Fig. 14 The DNN architecture

The overall sensor signal dataset and the partitioned signal dataset were adopted to execute the prediction on the CNN-CNN, LSTM-LSTM and DNN-DNN respectively. Taking the dataset C1C4 as the example, the prediction results are presented in Fig. 15 (un-partitioned signals) and Fig. 16 (partitioned signals).

The integrated result above of the dataset C1C4 is obtained after the prediction based on the corresponding sub-dataset of the tool wear initial, steady and severe stage. Figure 17 shows several representative fitting results of the individual sub-datasets, which covers different dataset, wear stage and deep learning model, and the prediction precision of the three ensemble models on the partitioned sub-datasets and the overall datasets is summarized in Table 4.

By comparing the results of Tables 3 and 4, it is obvious that, even if the un-partitioned datasets were adopted as the input, the prediction accuracy of the CNN-LSTM algorithm was higher than those of other deep learning algorithms. That is, the average accuracy of the CNN-LSTM algorithm is 75.6%, while those of the CNN-CNN, LSTM-LSTM and DNN-DNN algorithms were 71.3%, 65.8% and 67.4%, respectively. It is expected, since the proposed hybrid model integrates the advantages of the automatic feature extraction of CNN and the time series data sequence learning of LSTM. Moreover, from the perspective of employing the partitioned dataset, the accuracies of the four algorithms were all improved to some extents. The CNN-LSTM algorithm achieved the best average performance of 87.3%, followed by the CNN-CNN algorithm of 78.1%. Such a result proves again that, the proposed data partition strategy of the Hurst exponent is effective, which the segmented sensor data is beneficial for optimizing the prediction.

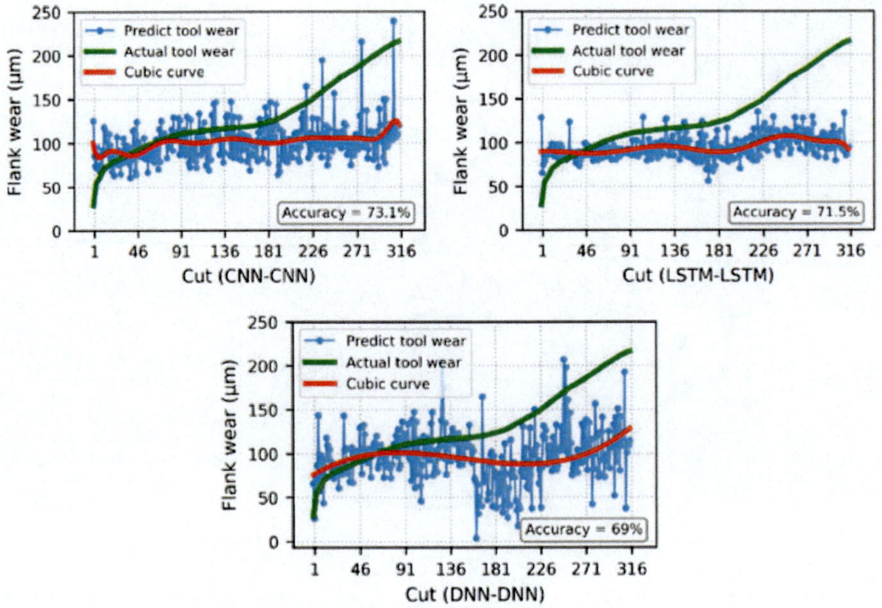

Fig. 15 Prediction using the un-partitioned dataset C1C4

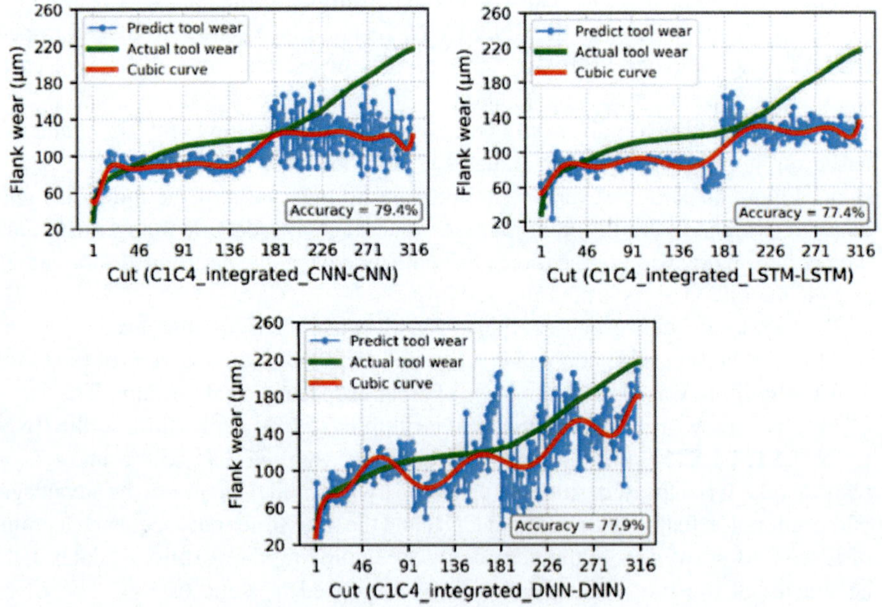

Fig. 16 Prediction using the partitioned dataset C1C4

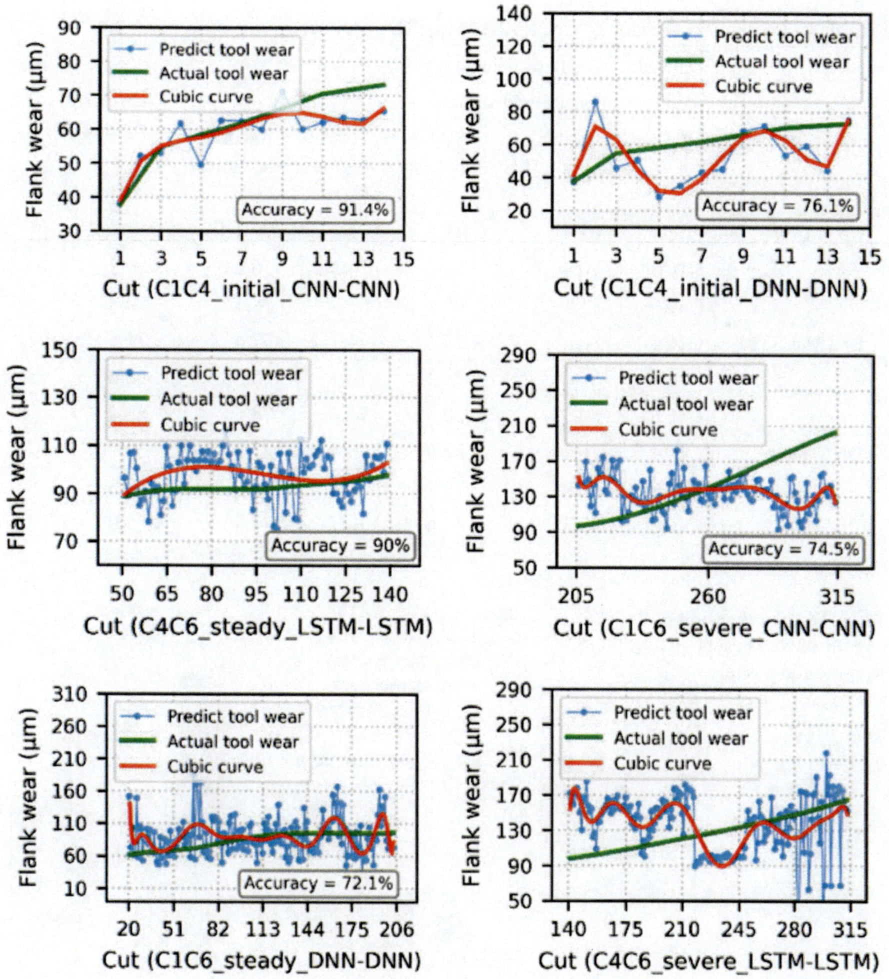

Fig. 17 Curve fitting of the partitioned datasets on prediction models

Computing efficiency is also a vital factor to investigate. In this research, the cumulative calculations of the computing time cost were carried out for the deep learning algorithms. The time usage of each prediction model to compute un-partitioned datasets and partitioned datasets are shown in Fig. 18.

From the aspect of model training time consumption, it can be clearly seen that whether the un-partitioned or the partitioned data was used as the input for training, the proposed CNN-LSTM consumes the minimum computing time among all four algorithms. The average time for the CNN-LSTM algorithm was approximate 0.33 h, the average time cost of the CNN-CNN model was the closest to the CNN-LSTM algorithm, about 0.52 h. The DNN-DNN and LSTM-LSTM algorithms consume 80% and 67% more time than the CNN-LSTM algorithm, respectively. In view of the CNN-LSTM model being outstanding in the prediction precision and efficiency,

Table 4 Prediction precision of CNN, LSTM and DNN models

Model	Dataset	Prediction accuracy				
		Partitioned signals				Un-partitioned signals (%)
		Initial region (%)	Steady region (%)	Severe region (%)	Integrated accuracy (%)	
CNN-CNN	C1C4	91.4	71.6	81.2	79.4	73.1
	C1C6	70.4	79.5	74.5	76.9	70.2
	C4C6	81.1	87.1	67.7	78.1	70.8
LSTM-LSTM	C1C4	80	79.3	74.3	77.4	71.5
	C1C6	84	78.4	72.1	75.5	60.6
	C4C6	80.7	90	71.4	78	65.4
DNN-DNN	C1C4	76.1	81.9	73.4	77.9	69
	C1C6	76.8	72.1	70.3	71.6	69.5
	C4C6	72.7	83.6	70.7	74.5	63.9

Fig. 18 Curve fitting of the partitioned datasets on prediction models

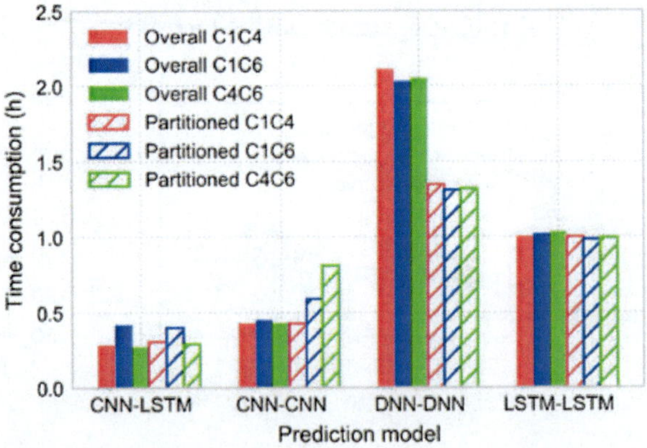

it can be considered a powerful and promising deep learning scheme for the cutting tool RUL regression prediction.

4.4 Comparison of the Hurst Exponent with Other Methods

Furthermore, to effectively investigate the adaptability of the Hurst exponent-based signal partition method, the performance comparison between the proposed method and other prevalent signal processing approaches were executed. These approaches are domain-feature extraction and PCA dimensionality reduction.

The signal processing of the prediction task is responsible to eliminate unwanted data from massive sensor signals, to improve the efficiency and accuracy of the prediction model with lower volume and valuable data. For the domain-feature extraction, the features of the signals adopted in this chapter are extracted in the time domain, frequency domain and time–frequency domain. In the time domain, statistical features are usually extracted to reflect the change of signal properties over time. And, as another indispensable feature extraction method, the frequency domain signal is able to present the signal change rate and provide the spectral feature for hidden information [28]. The fast Fourier transform (FFT) is the most common frequency domain transformation. It is used to convert the sensor signals of each cut cycle in this chapter. The FFT of the x-axis vibration signal of the 1st cut of the cutting tool C1 is shown in Fig. 19. Moreover, facing the nonlinear sensor signal, the time–frequency domain signal outperforms the methods to process signals in the time domain and frequency domain by capturing useful features, since it is conducive to explore the feature in transient and localized components. In this chapter, the Continuous Wavelet transform (CWT) is applied to transform each sensor signal to the time–frequency domain, which is a powerful method to represent the sensor signal into a two-dimensional plane. The CWT result of vibration signals of the 1st cut for C1 in the X-axis is shown in Fig. 20.

After the conversion of the frequency domain and time–frequency domain for sensor signals of the 315 cuts of each cutting tool, including the time domain, several widely employed features are extracted in every domain. It is noteworthy that the feature extraction in the time–frequency domain is based on a different scale, which is divided according to the sampling rate of the adopted dataset. Table 5 summarizes the extracted features of three domains in this chapter.

The validation dataset in this chapter contains 7 sensor signals for the 315 cut cycles of 3 cutting tools. Finally, there are 158,760 features obtained from raw sensor signal by the above feature extraction mentioned. Then, the generated features of each cutting tool are gathered as a new input dataset, and they are adopted to perform the flank wear prediction on the CNN-LSTM algorithm, the fitting result is shown in Fig. 21.

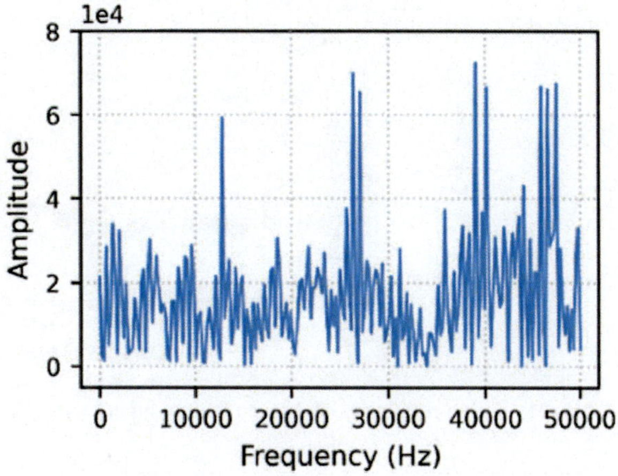

Fig. 19 FFT spectrum of the x-axis vibration signal of 1st cut of the cutting tool C1

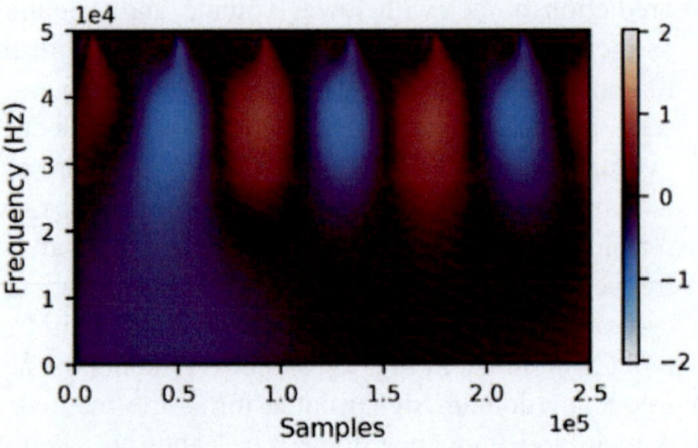

Fig. 20 The CWT plot of the vibration signal of the 1st cutting for the cutting tool C1

Table 5 Extracted feature from the different domain

Domain			
Time domain	Frequency domain	Time–frequency domain	
Mean Standard deviation Skewness Kurtosis Maximum Peak to peak	Frequency center Median frequency Root variance frequency	0–10,000 Hz 10,000–20,000 Hz 20,000–30,000 Hz 30,000–40,000 Hz 40,000–50,000 Hz	Mean Standard deviation Variance

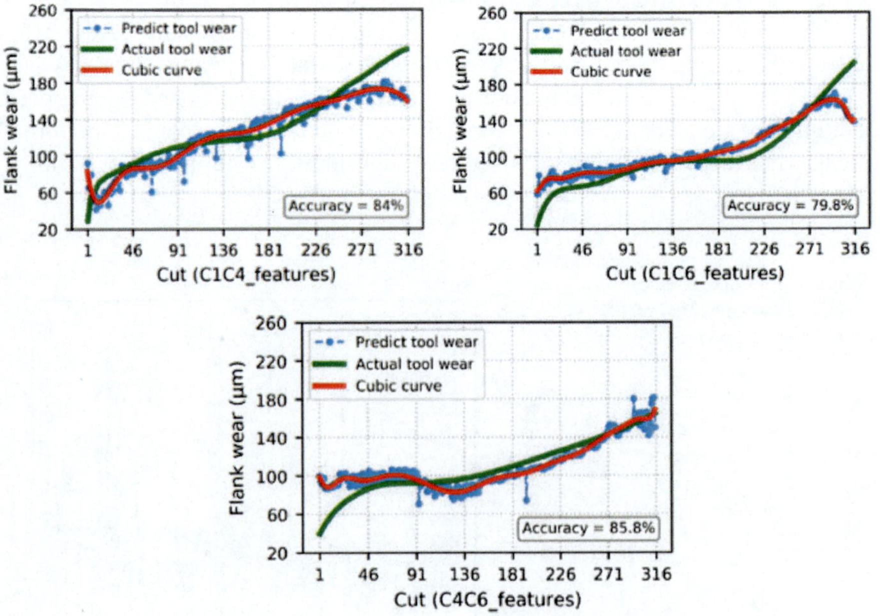

Fig. 21 Fitting results of the domain feature input by the CNN-LSTM algorithm

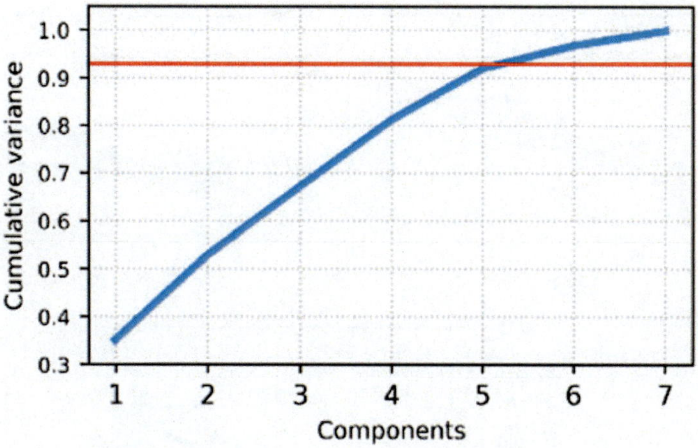

Fig. 22 The PCA cumulative variance plot for the 1st cutting of the cutting tool C1

Besides the domain-feature extraction, the signal processing based on the PCA is a prevailing approach for dimensionality reduction of the multi-sensor signal. It is capable to compress the signal dimension without discarding main information by mapping the raw sensor signal into a feature space with representative components. For the dimensionality reduction of the sensor signal by PCA, the number of components is necessary to be determined initially via the cumulative sum of explained variances, which provides the proportion of the retain information of the different number of components. Taking cutting tool C1 as an example, the cumulative sum of explained variances of the sensor signal in the 1st cut is shown in Fig. 22.

From the above figure, it is observed that five components generated by the PCA could remain over 90% original sensor signal information, which is the same outcome of the other two cutting tools. Thus, these five components are adopted to replace the raw sensor signals of every cutting tool, each component includes the same number of data as the original sensor signal. Then, the flank wear prediction based on the five components was executed by the CNN-LSTM algorithm. The input strategy is pair-wise concatenated as well. Figure 23 shows the prediction result.

Based on the described approach of the domain feature extraction and the PCA dimensionality reduction, the corresponding flank wear prediction result on the hybrid CNN-LSTM model are detailed in Table 6. Figure 24 shows the total processing time consumption of the sensor signals of the three datasets using these two methods and the proposed Hurst exponent method, as well as the cumulative time for their subsequent predictions.

By comparing the result of Tables 3 and 6, it is apparent that the prediction accuracy for tool wear based on the dataset partitioned through the proposed Hurst exponent was higher than applying the domain feature extraction and the PCA dimensionality reduction. Separately, the feature extraction in the three domains achieved the data volume decreasing to a great extent. Due to the elimination of redundant signals, its precision was closer to the Hurst exponent-based partition method and the model

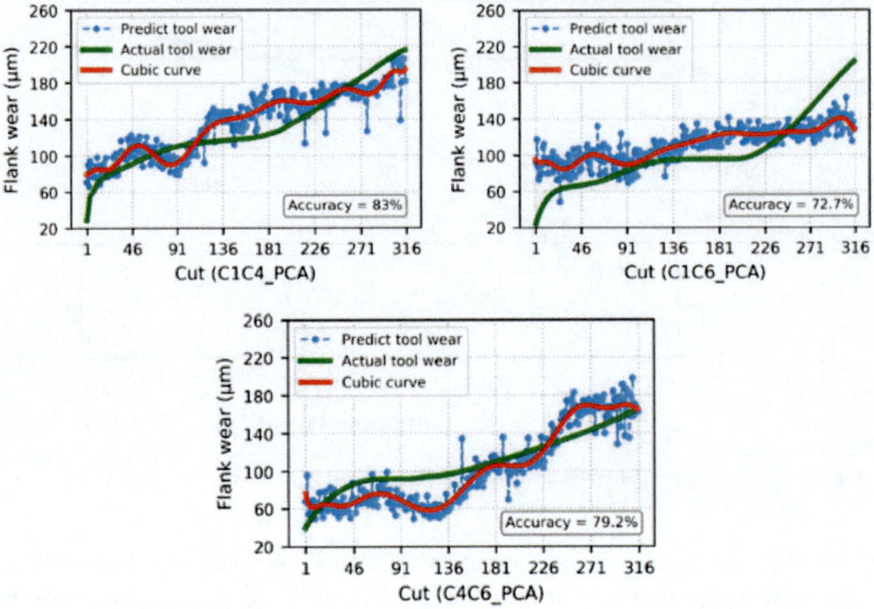

Fig. 23 The fitting results of the PCA component input on the CNN-LSTM algorithm

Table 6 Prediction result of the different data processing methods

Data processing method	Input dataset	Prediction accuracy (%)	Average accuracy (%)
Domain feature extraction	C1C4	84	83.2
	C1C6	79.8	
	C4C6	85.8	
PCA dimensionality reduction	C1C4	83	78.3
	C1C6	72.7	
	C4C6	79.2	

Fig. 24 Time consumption of different data processing methods

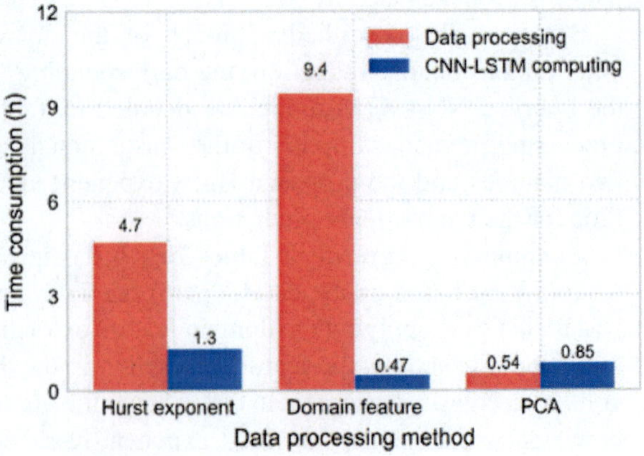

calculation time was also compressed greatly. However, the domain conversion and feature extraction consumed a large amount of time, which were 50% and 94% higher than those of the Hurst exponent method and the PCA dimensionality reduction, respectively. It is an obvious obstacle to practical applications. On the other hand, PCA showed the benefit of the computational time–cost saving, by condensing the raw sensor signal into five new representative components, an approximate 30% dimensionality reduction was implemented. Nevertheless, the prediction accuracy is sacrificed to reach this dimensionality reduction, and it is foreseen that the optimal components of the PCA will appear more deviation along with the increasing of the signal volume. According to these assessments, the introduced Hurst exponent-based signal partition displays a stronger ability than the other two prevalent data processing approaches. Its effectiveness for the flank wear prediction could be illustrated as that the Hurst exponent partition simplifies the system frame with non-complex computation, and helps accomplish the satisfying performance of the deep learning algorithm on the premise of not discarding the signal information.

4.5 *RUL Prediction*

After determining the superiority of the Hurst exponent partition method for the flank wear prediction, as the most ordinarily used tool life criteria, the flank wear further defined the cutting tool RUL based on the above prediction results. In general, the threshold of tool wear and tool life are determined by the actual application scenario and demand. Based on the dataset employed in this chapter, the life of each cutting tool is defined as 315 cycles. In order to intuitively display the remaining life of the tool, the 315 corresponding actual flank wear values are expressed linearly by the remaining processable cycle of the tool. Furthermore, the remaining processable cycle of the predicted flank wear can be calculated as follows:

$$RULpi = \frac{{w_p}^i \cdot {RUL_a}^i}{{w_a}^i} \tag{19}$$

where, RUL XE "RUL" a denotes the assigned actual remaining processable cycle, which is $\{315, 314, \cdots, 1\}$. w_p denotes the predicted flank wear, w_a denotes the actual measured flank wear, i denotes the number of the cuts.

A polynomial regression fitting model is then constructed based on the sample data obtained by Eq. (19) and the predicted flank wear for prediction of the RUL of the cutting tools. The regression function can be described as:

$$F(w) = \sum\nolimits_{j=0}^{k} b_j \cdot w^k \tag{20}$$

where, F denotes the RUL of a cutting tool, w denotes the flank wear, b_j denotes the regression coefficients.

Figure 25 shows the comparison of the RUL prediction results of the partitioned and the un-partitioned datasets. The RUL polynomial regression result of the partitioned datasets is depicted in Fig. 26.

It can be observed from Fig. 25 that the predicted value of the RUL fluctuates around the actual value, and the overall trend remains unchanged.

From the beginning to the end of the cutting process, the predicted value always fluctuates above or below the actual value, the relative error is large at the beginning of the machining, and the predicted value of the tool life is closer to the actual value during the subsequent cutting.

With the tool wear deteriorates, the remaining machining cycle get easier to estimate. From the perspective of the segmented dataset C1C4, it can be seen that between

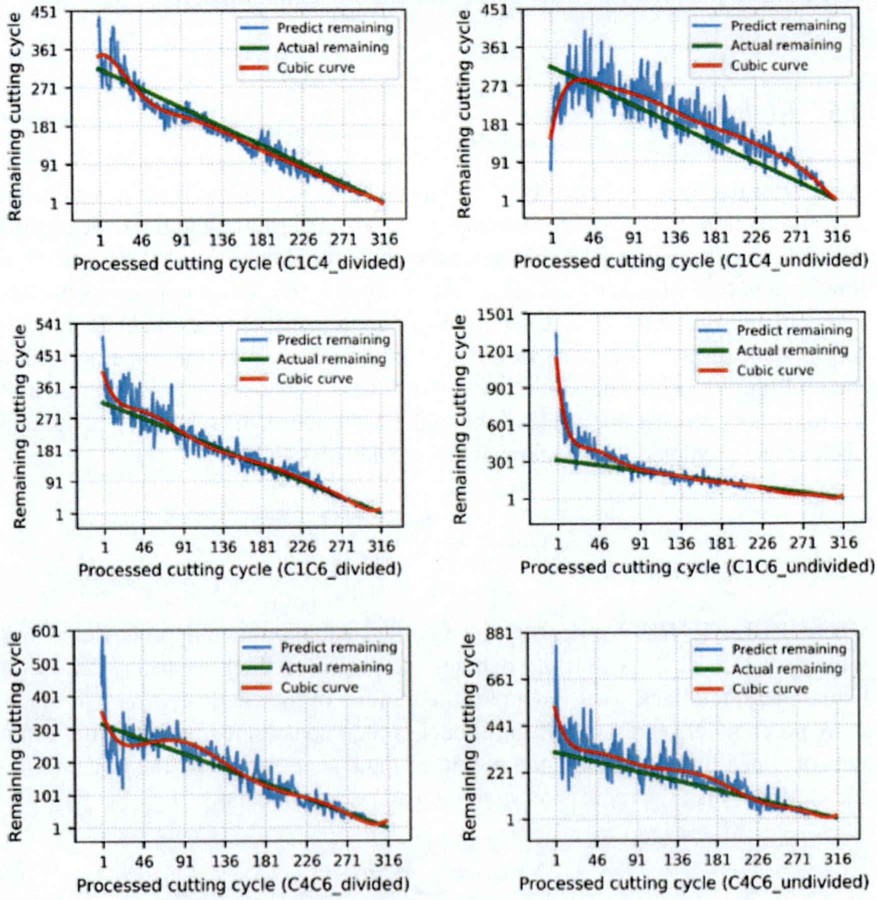

Fig. 25 Predicted RUL and actual RUL of 3 cutting tools using partitioned and un-partitioned datasets

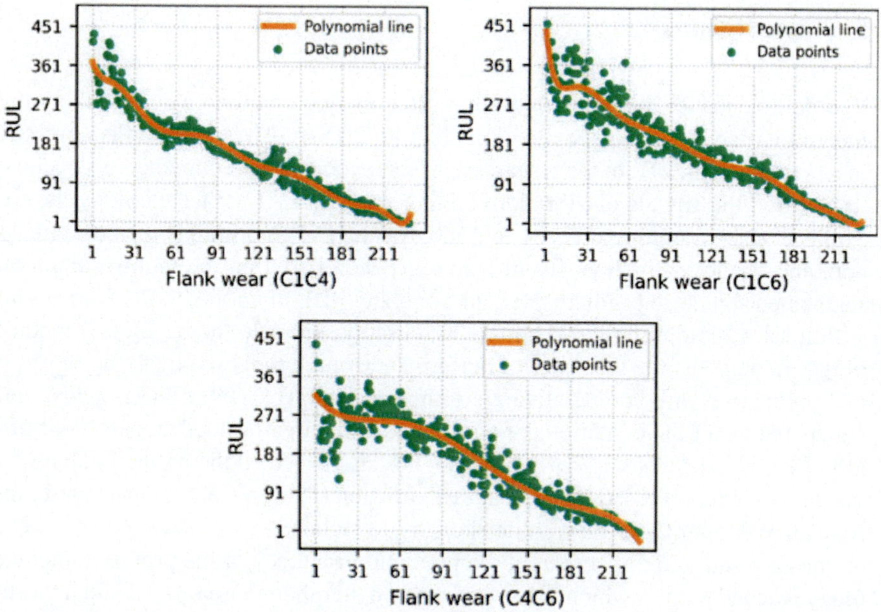

Fig. 26 Polynomial regression of the RUL of 3 cutting tools using partitioned datasets

the 1st cut and the 91th cut, the predicted cutting tool RUL fluctuates greatly and has certain errors. However, as the cutting continues until reaching the end of its life cycle (the 315th cutting cycles), the predicted value begins to approach the real value, and the prediction accuracy is guaranteed. In contrast, the prediction error of the un-partitioned dataset is still huge, in specific, the RUL of the cutting tool is underestimated in the interval from the 1st cut to the 20th cut, which may lead to early replacement of the cutting tool, and results in undesired waste.

After the 20th cut, the remaining life of the tool has been overestimated, which will cause the poor quality of the work-piece surface. Moreover, the prediction error of the undivided C1C6 dataset is more obvious that reaches the highest of 74%, and the undivided C4C6 dataset shows the overestimation in a long duration.

Additionally, the polynomial regression in Fig. 26 displays that the model effectively fits the flank wear and the RUL, it can prove that the model has excellent performance in RUL prediction. The accuracy of each regression prediction is assessed via the coefficient of determination (R^2), which evaluates the extent of the model interpret and predict the result, the R^2 value of 0.893, 0.898 and 0.856 are achieved by dataset C1C4, C1C6 and C4C6, respectively.

In the view of all results displayed above, the prediction accuracy of the dataset that partitioned by Hurst exponent presents significant improvement than that of the unsegmented dataset, and it illustrates the cutting tool RUL prediction system proposed in this chapter has the potential and superior performance.

5 Conclusions

In order to enhance the prediction of the RUL of a cutting tool, a new methodology integrating the Hurst exponent and the hybrid CNN-LSTM algorithm is developed in a systematic means. In the research, a Hurst exponent-based method is developed to partition the signals of vibration, cutting force and AE collected along the life-cycles of a set of cutting tools. A hybrid CNN-LSTM algorithm is then designed to combine feature extraction, fusion and regression based on the multi-sourced and segmented signals to estimate the flank wear and RUL of cutting tools. A case study with a set of complex sensor signals was used to validate the developed methodology. To demonstrate the superior performance of the proposed methodology, a set of benchmarks with comparative algorithms, including CNN, LSTM, DNN, were conducted under the conditions of partitioned and un-partitioned signals. Additionally, the proposed CNN-LSTM algorithm has also executed the prediction based on the dataset that processed by the feature extraction (in the time, frequency and time-frequency domain) and PCA dimensionally reduction. Results showed that, based on the case study in this research, the prediction accuracy of the proposed methodology reached 87.3%, which are significantly better than those of the benchmarking algorithms. Analyses on the results and observations were given in detail.

Future work will concentrate on the applications of the methodology on more case studies to prove the improved performance and generality. It will also investigate the possibility of further enhancing the accuracy of prediction, such as de-noising on signals and hybridization with effective strategies.

References:

1. Yousefi S, Zohoor M (2019) Effect of cutting parameters on the dimensional accuracy and surface finish in the hard turning of MDN250 steel with cubic boron nitride tool, for developing a knowledge base expert system. Int J Mech Mater Eng 14(1)
2. Zhou Y, Xue W (2018) Review of tool condition monitoring methods in milling processes. Int Adv Manufact Technol 96: 2509–2523
3. Serin G., Sener B., Ozbayoglu A.M., Unver H.O., 2020. Review of tool condition monitoring in machining and opportunities for deep learning. Int J Adv Manufact Technol. 109:953–947
4. Kaya B, Oysu C, Ertunc H, Ocak H (2012) A support vector machine-based online tool condition monitoring for milling using sensor fusion and a genetic algorithm. Proc Inst Mech Eng, Part B: J Eng Manufact 226(11):1808–1818
5. Shankar S., Mohanraj T., Pramanik A., 2019. Tool condition monitoring while using vegetable based cutting fluids during milling of Inconel 625. J Adv Manufact Syst 18(4):563–581
6. Rajamani D, Esakki B, Arunkumar P, Velu R (2018) Fuzzy logic-based expert system for prediction of wear rate in selective inhibition sintered HDPE parts. Mater Today: Proc 5(2):6072–6081
7. Zhang C, Yao X, Zhang J, Jin H (2016) Tool condition monitoring and remaining useful life prognostic based on a wireless sensor in dry milling operations. Sensors 16(6):795
8. Yu J, Liang S, Tang D, Liu H (2016) A weighted hidden Markov model approach for continuous-state tool wear monitoring and tool life prediction. Int J Adv Manufact Technol 91(1–4):201–211

9. Wu J, Su Y, Cheng Y, Shao X, Deng C, Liu C (2018) Multi-sensor information fusion for remaining useful life prediction of machining tools by adaptive network based fuzzy inference system. Appl Soft Comput 68:13–23
10. Yang Y et al (2019) Research on the milling tool wear and life prediction by establishing an integrated predictive model. Measurement 145:178–189
11. Wu X, Liu Y, Zhou X, Mou A (2019) Automatic identification of tool wear based on convolutional neural network in face milling process. Sensors 19(18):3817
12. Zang H, Liu L, Sun L, Cheng L, Wei Z, Sun G (2020) Short-term global horizontal irradiance forecasting based on a hybrid CNN XE "CNN" -LSTM XE "LSTM" model with spatiotemporal correlations. Renew Energy 160:26–41
13. Zhu J, Chen H, Ye W (2020) A hybrid CNN XE "CNN" –LSTM XE "LSTM" network for the classification of human activities based on micro-doppler Radar. IEEE Access 8:24713–24720
14. Bogaerts T, Masegosa A, Angarita-Zapata J, Onieva E, Hellinckx P (2020) A graph CNN XE "CNN" -LSTM XE "LSTM" neural network for short and long-term traffic forecasting based on trajectory data. Trans Res Part C: Emerg Technol 112:62–77
15. Kim T, Cho S (2019) Predicting residential energy consumption using CNN XE "CNN" -LSTM XE "LSTM" neural networks. Energy 182:72–81
16. He Y et al (2019) Application of CNN XE "CNN" -LSTM XE "LSTM" in gradual changing fault diagnosis of rod pumping system. Math Problem Eng 2019:1–9
17. Low L, Yan S, Kwan Y, Tan C, Thumboo J (2018) Assessing the validity of a data driven segmentation approach: A 4 year longitudinal study of healthcare utilization and mortality. PLoS ONE 13(4):e0195243
18. Opałka S, Stasiak B, Szajerman D, Wojciechowski A (2018) Multi-channel convolutional neural networks architecture feeding for effective EEG mental tasks classification. Sensors 18(10):3451
19. Knight M, Nunes M (2018) Long memory estimation for complex-valued time series. Stat Comput 29(3):517–536
20. Guan S, Pang H, Song W, Kang Z (2018) Cutting tool wear recognition based on MF-DFA feature and LS-SVM algorithm. Trans Chin Soc Agricult Eng 34(14):61–68
21. Mohanty S, Gupta K, Raju K (2018) Hurst based vibro-acoustic feature extraction of bearing using EMD and VMD. Measurement 117:200–220
22. Lotfalinezhad H, Maleki A (2020) TTA, a new approach to estimate Hurst exponent with less estimation error and computational time. Phys A 553:124093
23. Fawaz HI, Forestier G, Weber J, Idoumghar L, Muller P (2019) Deep learning XE "Deep learning" for time series classification: A review. Data Min Knowl Disc 33(4):917–963
24. Yamashita R, Nishio M, Do R, Togashi K (2018) Convolutional neural networks: an overview and application in radiology. Insights Into Imaging 9(4):611–629
25. Martínez-Arellano G, Terrazas G, Ratchev S (2019) Tool wear classification using time series imaging and deep learning. Int J Adv Manuf Technol 104(9–12):3647–3662
26. Rusinek R, Borowiec M (2015) Stability analysis of titanium alloy milling by multiscale entropy and Hurst exponent. Eur Phys J Plus 130(10)
27. Cheng H, Chen H, Li Z, Cheng X (2020) Ensemble 1-D CNN XE "CNN" diagnosis model for VRF system refrigerant charge faults under heating condition. Energy Buil 224:110256
28. Herff C, Krusienski D (2018) Extracting features from time series. Fundamentals of Clinical Data Science. pp 85–100.

Thermal Error Prediction for Heavy-Duty CNC Machines Enabled by Long Short-Term Memory Networks and Fog-Cloud Architecture

Y. C. Liang, W. D. Li, P. Lou, and J. M. Hu

1 Introduction

Heavy-duty CNC machines are highly demanded in some important applications for manufacturing large-scale and high-end products, such as steam turbines, large nuclear pumps, marine propellers, and large aircraft wings [1]. For machining processes conducted on heavy-duty CNC machines, precision is an essential technical requirement. It is paramount to develop effective error prediction modelling and then compensation technologies to minimize machining errors. The thermal error is a principal one accounting for about 40–70% of the total machining errors [2]. It is caused by heat and temperature rising during machining processes, leading to deformation of machine elements and further machining inaccuracy. There are mainly two types of heat sources generating thermal errors: (1) internal sources—heat is generated from cutting, friction and moving processes of machine elements; (2) external sources—heat is related to surrounding environmental temperature variations. To improve machining accuracy for heavy-duty CNC machines, it is prominent to develop an effective model for thermal error prediction, which can represent relationships between heat sources and the impact on thermal errors. Based on the model, an appropriate compensation strategy can be further developed to enhance the overall machining accuracy.

Y. C. Liang · W. D. Li (✉)
Faculty of Engineering, Environment and Computing, Coventry University, Coventry CV1 5FB, UK
e-mail: weidong.li@coventry.ac.uk

W. D. Li
School of Logistics Engineering, Wuhan University of Technology, Wuhan 430070, China

P. Lou · J. M. Hu
School of Information Engineering, Wuhan University of Technology, Wuhan 430070, China

W. Li et al. (eds.), *Data Driven Smart Manufacturing Technologies and Applications*, Springer Series in Advanced Manufacturing,
https://doi.org/10.1007/978-3-030-66849-5_6

In the past, physics-based modelling has been actively investigated to predict thermal errors for CNC machine systems [3]. In the modelling process, physical mechanisms of machining elements are established. Finite Element Method (FEM), Finite Difference Method (FDM), or Finite Difference Element Method (FDEM), is then applied to establish thermal deformation fields of machine elements. However, due to the large-scale structures and dynamic operation environments of heavy-duty CNC machines, it is challenging to developing accurate physical mechanisms for analyzing and predicting thermal errors preciously. In recent years, data-based modelling has been actively explored as an increasingly popular solution for thermal error prediction. By leveraging the latest deep learning technologies, such as Convolutional Neural Networks (CNNs) and Recurrent Neural Networks (RNNs), thermal error related features can be mined from a large volume of thermal data collected from machine elements through deployed sensors. In comparison with physics-based modelling, data-based modelling is more flexible and accurate in supporting the machining processes conducted through heavy-duty CNC machines.

However, data-based modelling could be severely hindered by the less efficient processes of collecting and handling large-volume thermal data. To mitigate the challenge, this chapter presents an improved data-driven system for thermal error prediction on heavy-duty CNC machines. Innovative characteristics of the research are given below:

- A large number of thermal sensors is required to install into the large-scale structure of a heavy-duty machine in order to collect sufficient data for accurate thermal modelling. Nevertheless, sensors could be excessively deployed to generate redundant data if without an appropriate installation guidance provided. To optimize the number of deployed sensors and minimize excessive data, this research is innovative in integrating physics-based modelling to facilitate data-based modelling. That is, FEA is conducted based on the structure of a heavy-duty CNC machine and its heat sources. The results of FEA, which show the heat distributions over the entire structure of the heavy-duty machine, provide sensible instructions to minimize unnecessary sensors and data;
- Meanwhile, it is essential to design a more effective deep learning algorithm for predicting thermal errors on machines. The LSTM networks, which is an improved RNNs, is robust for time-series data processing and relatively insensitive to unknown gap lengths in the data. Based on the characteristic, in this research, a LSTM networks is designed to build the relationship between key heat sources and the thermal deformation fields of machine elements for thermal error prediction. This LSTM design is justified by benchmarking with different LSTM designs and other intelligent algorithms. Furthermore, to reduce data collinearity and computational workloads to better support the LSTM, an improved Grey Relational Analysis (*i*GRA) is developed to pre-process thermal data;
- A cloud architecture with associated computational resources on a cloud server is a popular solution to support data-based modelling. However, the efficiency of communicating monitored data could be seriously affected by the limited bandwidth and high latency of the industrial Internet. In this research, a fog-cloud

architecture based on the "divide-and-conquer" strategy is designed to tackle the issue. In the new architecture, *i*GRA-based data pre-processing and LSTM-based prediction modelling are conducted locally on a fog layer to expedite decision making. FEA and LSTM training, which are computationally intensive activities, are processed on a cloud layer to leverage its computational resources. Quantitative analyses are conducted to showcase the advantage of the fog-cloud design for thermal error prediction of heavy-duty machines.

In this chapter, the developed system was validated using a heavy-duty CNC machine namely ZK5540, which was made by the Wuhan Heavy Duty Machine Tool Group Corporation. Industrial case studies demonstrated that, by adopting the system, the volume of transmitted data was minimized by 52.63% and the thermal errors were reduced by 46.53%, in comparison with processes without using the developed system. Based on the case studies and benchmarking analyses, the benefit of adopting the system in terms of data processing efficiency and machining accuracy improvement is clearly exhibited.

2 Related Works

2.1 Data-Based Modelling for Thermal Error Prediction

According to the survey [4], data-based modelling for thermal error prediction has been becoming an increasingly popular approach. There are usually two main steps to build a data-based thermal model: (1) to optimize deployment points of sensors, and (2) to develop thermal error prediction models. Miao et al. [5] developed a principal component regression algorithm to remove data points with collinearity relationships. Cheng et al. [6] proposed a grey system theory to analyze the data similarity among temperature sensors. Based on the analysis, sensors mostly sensitive to heat were selected. Case studies demonstrated the temperature sensors were decreased from 24 to 7 after this analysis. Abdulshahed et al. [7] proposed a method based on grey modelling and fuzzy c-means clustering to determine key temperature points on machines. 525 discrete spots were classified into 8 groups for further modelling and data minimization. Liu et al. [8] designed an optimal selection method of temperature-sensitive measuring points. In the method, the degree of temperature sensitivity was defined and used to select measuring points with high sensitivity to thermal error. The selected points were then classified with fuzzy clustering and grey correlation grades, and temperature-sensitive measuring points were selected based on the analysis of temperature sensor locations. Results showed that the number of the measuring points was reduced from 27 to 5 by using the method. Ma et al. [9] used the genetic algorithm and particle swarm optimization to optimize the parameters of an artificial neural networks to build up a thermal error prediction model effectively in the aspects of accuracy, convergence performance and robustness. Li et al. [10] devised a clustering method to select the most suitable sensor points for data analysis.

Li et al. [11] developed a thermal error prediction model based on a hybrid particle swarm optimization and artificial neural network. Temperature measurement points were clustered by a SOM neural network, and analysis was conducted to explore the correlation between the thermal sensitive points and the thermal error. Fujishima et al. [12] developed a novel thermal displacement compensation method using a deep learning algorithm. In the algorithm, reliability of thermal displacement prediction was evaluated and compensation weights were adjusted adaptively. Yao et al. [13] designed an optimal composite model (OM) for modelling spindle thermal error prediction. The grey model and the least squares support vector machine (LS-SVM) were used to establish the thermal error prediction model. Then, the OM model was used to adjust the weighting coefficients of LS-SVM to fine-tune the prediction model based on practical thermal error data. The above works are also summarized in Table 1.

The aforementioned methods for sensor point selection are based on correlation analysis on collected data to reduce redundant sensors. On the other hand, for heavy-duty CNC machines, it is challenging to deciding initially deployed sensor points to create a suitable initial dataset for further optimization. Trials are usually expensive and therefore a guidance on initially deployed points is required. There are limited research conducted in this area.

Table 1 Some research works of data-based modelling for thermal error prediction

Research	Functions	Intelligent algorithms
Miao et al. (2015) [5]	To remove data points with collinearity relationship	Principal component regression algorithm
Cheng et al. (2015) [6]	To remove data points by analysing data similarity among temperature sensors	Grey system theory
Abdulshahed et al. (2015) [7]	To determine key temperature points	Grey modelling and fuzzy c-means clustering
Liu et al. (2015) [8]	To remove data points by analysing data similarity among temperature sensors	Fuzzy clustering and grey correlation grade
Ma et al. (2016) [9]	Thermal error prediction modelling	Genetic algorithm, particle swarm optimisation, artificial neural networks
Li et al. (2017) [10]	To select the most suitable sensor points for data analysis	Fuzzy clustering
Li et al. (2019) [11]	Thermal error prediction modelling	Hybrid particle swarm optimisation and artificial neural networks
Fujishima et al. (2019) [12]	Thermal displacement compensation	A deep learning algorithm
Yao et al. (2020) [13]	Thermal error prediction modelling	Grey model and the least squares support vector machine, composite model

2.2 STM Networks for Data-Based Manufacturing Applications

Owing to the advantage of effectively solving the common problem of "vanishing/exploding gradients" in deep learning and revealing patterns for time-series data, the LSTM networks for data-based manufacturing applications has been widely researched [14]. Zhang et al. [15] designed a LSTM networks for machine remaining life prediction. Li et al. [16] proposed a Quantum weighted LSTM networks to predict degradation trend of rotating machinery. In the research, the minimum prediction error using the LSTM networks proved to be 1.54% while other comparative algorithms such as RNNs and Support Vector Machine (SVM) were more than 1.90% in prediction error. Yang et al. [17] developed a LSTM networks to detect and isolate faults for electro-mechanical actuators. The method achieved accuracy for more than 99% and good robustness, while other algorithm such as SVM was 94.4% in accuracy. An et al. [18] designed data-driven modelling for remaining useful life prediction for machining cutters using a convolutional and stacked LSTM networks. The prediction accuracy of the proposed method was up to 90% in comparison with 86% accuracy using CNNs. Xia et al. [19] proposed an ensemble framework based on convolutional bi-directional LSTM networks to predict remaining useful life of machine tools, and the proposed method was tested with multiple datasets to demonstrate the accurate and robust prediction performance of the designed LSTM. The Root Mean Square Error (RMSE) can achieve 12.66, and other algorithms such as Deep Neural Networks (DNNs) and CNNs can achieve 19.05 and 15.50, respectively. Liu and Zhu [20] developed a two-stage approach for predicting the remaining useful life of machine tools utilizing a bidirectional LSTM networks. The developed method was compared with different algorithms such as Support Vector Machine (SVM) and decision tree for benchmarking. The proposed method achieved 4.18% in Mean Absolute Percentage Error (MAPE), which was the lowest compared to 7.16%, 6.29 and 7.08% of using the decision tree, SVM and boosted tree respectively.

The above research works have proved that LSTM exhibits good performance in comparison with other comparative algorithms especially in processing time-series data. It is therefore worth exploring how to design an appropriate LSTM networks to facilitate heavy-duty machine applications. Meanwhile, the data collinearity issue for input data would degrade the performance of LSTM, thereby inspiring design of a sensible strategy to further improve the performance of the LSTM networks for manufacturing applications.

2.3 Information Architecture for Data-Based Modelling

For data-based modelling to support heavy-duty CNC machines, thermal data are usually collected and transmitted to a cloud server for training and applying intelligent algorithms. However, based on the limited bandwidth of the industrial Internet in manufacturing companies, data transmission could be severely congested due to the heavy data traffic from sensors deployed on machine elements to a cloud server. To expedite system efficiency, a fog (From cOre to edGe) computing model has been actively to adopt a "divide-and-conquer" strategy in avoiding low latency of data transmission [21, 22]. A fog model consists of gateways with certain computing capacities to process monitored data. Only data that are necessary for central processing are transmitted to a cloud server. That is, the most computationally intensive jobs are undertaken on cloud while the fog model is responsible for local processing. Thus, the overall data transmission and computational performance can be significantly enhanced in a fog architecture [23]. Wu et al. [24] proposed a fog computing system to achieve process monitoring and prognosis for manufacturing company. The system is capable of collecting and processing large-scale data in real time. A machine learning algorithm was implemented to predict tool wear through CNC machining operations. Sood and Mahajan [24] proposed a fog system for the healthcare sector to distinguish, detect and prevent mosquito borne diseases (MBDs). The system can assimilate, analyze and share medical information among users and healthcare service providers.

For thermal error prediction on heavy-duty machines, there is no reported work yet to leverage the research progress of a fog-cloud architecture for improving data transmission and computing efficiency. Thus, in this research, investigations will be conducted on how to design an appropriate fog-cloud architecture to optimize this application.

3 Research Methodologies and System Design

3.1 FEA Analysis for Optimized Deployment of Sensors

In this research, a heavy-duty CNC machine, namely ZK5540, is used as an example. In the machine, thermal sensors and laser displacement sensors are installed to collect thermal data and spindle deformation data. The ZK5540 machine, deployed sensors, and data processing system are shown in Fig. 1.

The collected data are used to train a LSTM networks designed in this research to predict thermal errors for adaptive compensation on the fly. To provide a guidance on how to place thermal sensors more effectively, an FEA model is designed to analyze the thermal distribution of the machine. To build up the FEA model, the main internal heat sources of the machine are identified as thermal loads. They are primarily from: (1) the current of the stator and rotor coils of the spindle motor and the servo motor, (2) the friction of the bearings of the spindle motor and the servo motor, and (3) the friction of the guide rails. To calculate heat generated during the machining process, these aspects are formulated in detail as follows.

Heat is generated due to the current in the stator and rotor coils of the spindle motor and servo motor:

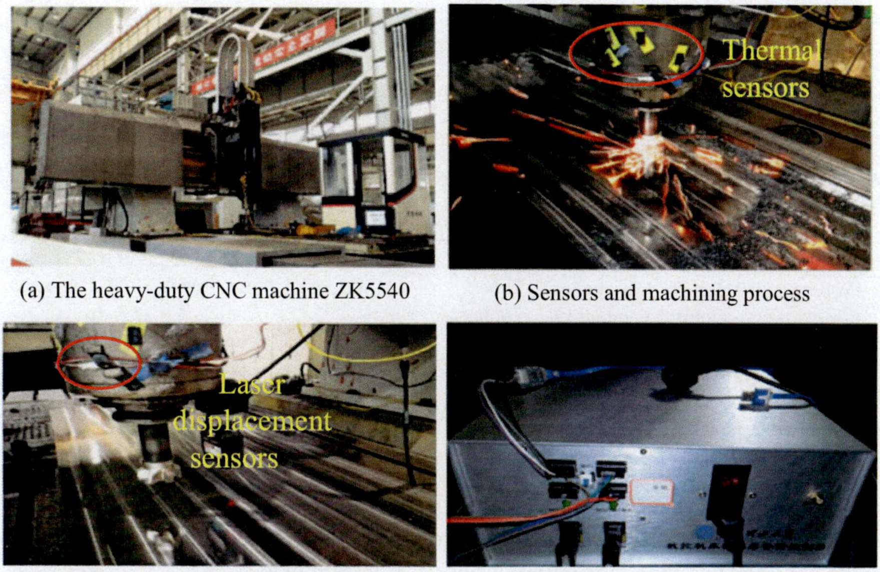

(a) The heavy-duty CNC machine ZK5540 (b) Sensors and machining process

(c) Sensors and machining process (d) Data processing system

Fig. 1 The heavy-duty CNC machining system and sensor deployment

$$P_e = I \cdot R^2 \tag{1}$$

where P_e is the thermal power due to the current; I is the current flowing through the stator and rotor coils; R is the resistance of the stator and rotor coils.

Meanwhile, heat is generated from the friction of the spindle and servo bearings of the spindle motor and servo motor:

$$P_f = 0.0001047 \cdot n \cdot M \tag{2}$$

where P_f is the thermal power generated due to the friction of the spindle bearing; n is the rotational speed of the spindle; M is the total friction torque.

At the same time, heat is also generated because of the friction on the guide rails:

$$P_m = \mu_f \cdot N \cdot V \tag{3}$$

where P_m is the thermal power due to friction on the guide rail; μ_f is the friction coefficient; N is the normal force; V is the velocity of translational travelling.

Table 2 shows important properties and parameters of the ZK5540 machine. To build its FEA model, meshing and material properties of the machine are applied. Meanwhile, thermal loads defined as above are set. Table 3 includes detailed meshing information.

Table 2 Properties of the heavy-duty machine

Property	Main material	Working table width	Working table length	Travelling range for axis X	Travelling range for axis Y	Travelling range for axis Z
Value	Steel	2000 mm	4500 mm	5000 mm	4000 mm	650 mm

Table 3 Meshing properties of the heavy-duty machine

Property	Tetra quantity	Edge quantity	Face quantity	Min Edge Angle (degrees)	Max Edge Angle (degrees)	Max Aspect Ratio
Value	87,329	139,752	199,783	0	173.98	12.51

Figure 2 shows the heat distribution of ZK5540 analysed by FEA, which is used to identify appropriate positions to deploy sensors (some deployments based on the FEA result are shown in Fig. 3). According to the FEA result and the monitoring areas on the machine, the following observations are made:

- Temperature primarily increases around the spindle at about 80 °C the highest. The main reason is that heat is primarily generated there due to the current and friction of motors. Heat is difficult for dissipation due to the limited area for conduction. Therefore, majority of the temperature sensors, i.e., 52 thermal sensors, are installed all-round the spindle.
- Temperature also increases to approximately 35 °C around the columns due to the current and friction of motors. The temperature is not as high as the spindle

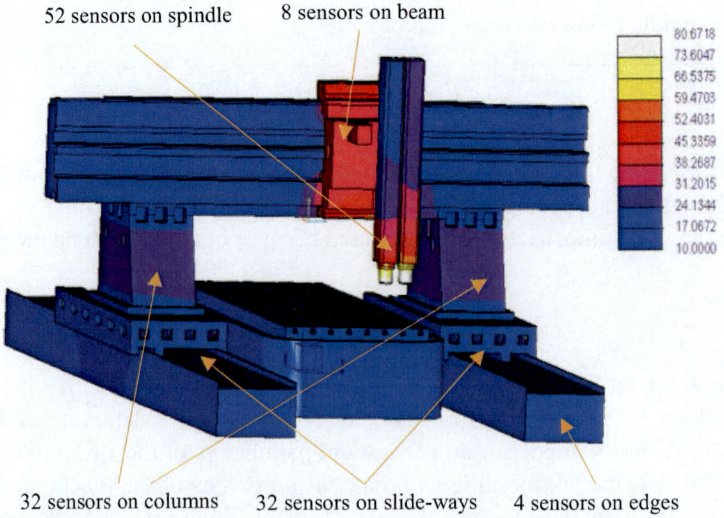

Fig. 2 The heat distribution analysed by FEA to guide the deployment positions for sensors

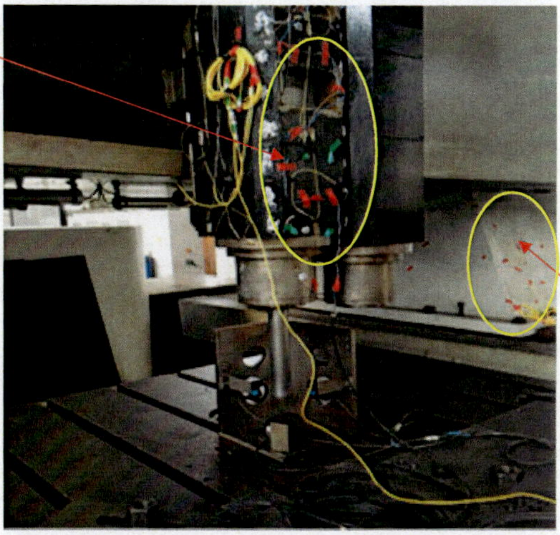

Fig. 3 Illustration of some deployed sensors into the machine

because it has a much greater area for heat dissipation. Thus, 32 sensors are installed on columns (16 sensors on each column).

- Heat could be generated around the slide-ways and beam due to friction. Similarly, the temperature is not as high as that of the spindle. 32 and 8 sensors are installed on the slide-ways and beams, respectively.
- To detect the environment temperature, 4 sensors are installed on the edges of the slide-ways. The temperature difference between the environment temperature and temperature on machining components, including the spindle, columns, slide-ways and beam, can be calculated.

3.2 Improved Grey Relational Analysis (iGRA)

To reduce data collinearity and processing workloads, collected thermal data are expected to be pre-processed. Grey Relational Analysis (GRA) is a useful approach to identify relationship coefficients among sensor nodes in order to remove duplication for data quality improvement. However, the GRA process might be not adaptive in practical applications considering that it is difficult to identify a reference sensor, which refers to a sensor that can reflects the temperature distribution with the highest sensitivity. To tackle the issue, an improved GRA, i.e., *i*GRA, is developed in this research. Some critical steps for *i*GRA are depicted below.

1. Normalization: t_{ij} represents a temperature measured by sensor node i at time j. All the measured temperatures are grouped as a set$T = \{t_{ij}\}(i = 1, 2, \ldots, n;\ j = 1, 2, \ldots, m)$, where n is the maximum sensor node index and m

is the maximum time index. The temperature t_{ij} is normalized to t'_{ij} to facilitate further processing. The normalization process is below:

$$t'_{ij} = \frac{t_{ij} - Min(T)}{Max(T) - Min(T)} \tag{4}$$

where $Max(T)$ and $Min(T)$ represent the maximum and minimum temperatures in the set T.

2. Calculation of the grey relational coefficient (including improvements over the GRA algorithm): The grey relational coefficient $g\left(t'_{rj}, t'_{ij}\right)$ can determine how close the temperature of a sensor t'_{ij} is to the temperature of the reference sensor t'_{rj}. Its computation is below:

$$g\left(t'_{rj}, t'_{ij}\right) = \frac{\Delta_{min} + \mu \cdot \Delta_{max}}{\Delta_{ij} + \mu \cdot \Delta_{max}} \tag{5}$$

where $\Delta_{ij} = \left|t'_{rj} - t'_{ij}\right|$; $\Delta_{min} = Min(\Delta_{ij})$; $\Delta_{max} = Max(\Delta_{ij})$; μ, which is within 0 and 1, is a distinguishing coefficient.
μ is designed to expand or compress the range of the grey relational coefficient. In this research, to improve the performance of grey relational analysis, μ is set 0.5 by default (Kuo et al., 2008).
In the GRA algorithm, only one reference sensor is set to be compared with other sensors for their temperatures. However, it is laborious to identify which sensor will be appropriate to be such a reference. Therefore, robustness is improved in the *i*GRA algorithm. That is, there will not be just one reference, and all the sensor data are selected as references for comparisons. Meanwhile, the following is designed for computational minimization in the *i*GRA algorithm:
As aforementioned, instead of setting up a fixed reference sensor, each sensor data is set as a reference along with the computation process. All the other data sequences are compared with the standard sequence according to the total grey relational grade $g\left(t'_{rj}, t'_{ij}\right)$ based on Eqs. (5 Chap. 5). For example, t'_{rj} is set as a standard data sequence and compared with t'_{ij}. The same result will be computed again when t'_{ij} is set as the standard data sequence and compared with t'_{rj}. To minimise the redundant computation, the same pair of data sequences will only be computed once. Figure 4 shows an example to illustrate the difference between the GRA and *i*GRA algorithms, in which each curved double arrow connector represents a grey relational coefficient between two sensors.
Therefore, the total number of grey relational coefficients to be calculated is:

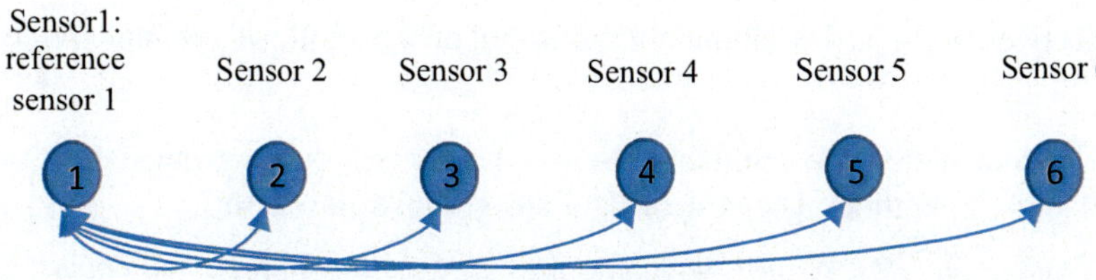

(a) Example of the GRA algorithm with 6 sensors.

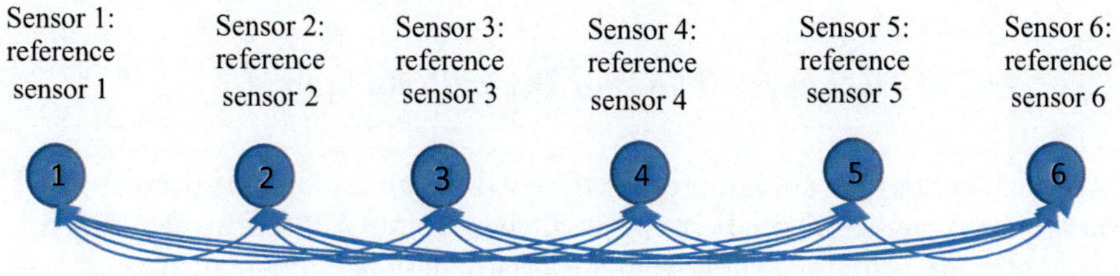

(b) Example of the *i*GRA algorithm with 6 sensors.

Fig. 4 Example of illustrating the GRA and *i*GRA algorithms using 6 sensors

$$N = \frac{n \times (n-1)}{2} \tag{6}$$

where N is total amount of grey relational coefficients to be calculated; n is the number of sensors.

3. Calculation of the total grey relational grade: The total grey relational grade $G(T_r', T_i')$ is calculated based on $g\left(t_{rj}', t_{ij}'\right)$ according to the following formula:

$$G\left(T_r', T_i'\right) = \sum_{j=1}^{m} \left(w_j \cdot g\left(t_{rj}', t_{ij}'\right)\right) \tag{7}$$

where w_j is a weight decided by a user; $\sum_{i=1}^{n} w_j = 1$.

4. Sensor selection: The above $G(T_r', T_i')$ is used to calculate the similarity between sensor nodes T_r' and T_i'. The higher $G(T_r', T_i')$, the closer T_r' and T_i'. If $G(T_r', T_i')$ is within a pre-defined threshold θ, T_i' can be removed for minimizing data volume and improving the training accuracy of the LSTM networks. That is, the threshold is decided when the following fitness achieves the highest value:

$$fitness = max(w_1 \cdot N_{data} + w_2 \cdot N_{accuracy}) \tag{8}$$

where N_{data} is the normalized saved data volume; $N_{accuracy}$ is the normalized training accuracy of the LSTM networks; w_1 and w_2 are the weights for the two objectives ($w_1 + w_2 = 1$); the normalization process follows the same principle as Eq. (4).

It is assumed that the number of sensor nodes is reduced from n to $n^{'}$ after this selection operation. The rest of data are grouped into a set$T^{'}_{j^{'}} = \left\{t^{'}_{ij^{'}}\right\}(i = 1, 2, \ldots, n; j^{'} = 1, 2, \ldots, m^{'})$, which are further used for training and applying the LSTM networks for thermal error prediction.

3.3 LSTM Design for Thermal Error Prediction

Thermal data, which are pre-processed by *i*GRA, are fed into the designed LSTM networks to predict thermal errors. As the base for a LSTM networks, RNNs is a deep learning neural networks with a specially designed topology, through which the connections among data points within arrays can be built by hidden states. Information from previous data points generates impact on the following prediction, so that the prediction can be improved compared with traditional neural networks [25]. Owing to the capability of learning 'context' within data arrays, RNNs has been widely used in time-series data prediction, such as stock market forecasting, language modelling, machine degradation prediction, etc. Figure 5 shows a typical structure of RNNs. To predict thermal errors by a RNNs, the vector in the hidden layer h_t and output vector E_t (thermal error at time t) can be calculated by the following equations [26]:

$$h_t = \sigma(W_h \cdot T^{''} + U_h \cdot h_{t-1} + b_h) \quad (9)$$

$$E_t = \sigma(W_E \cdot h_t + b_E) \quad (10)$$

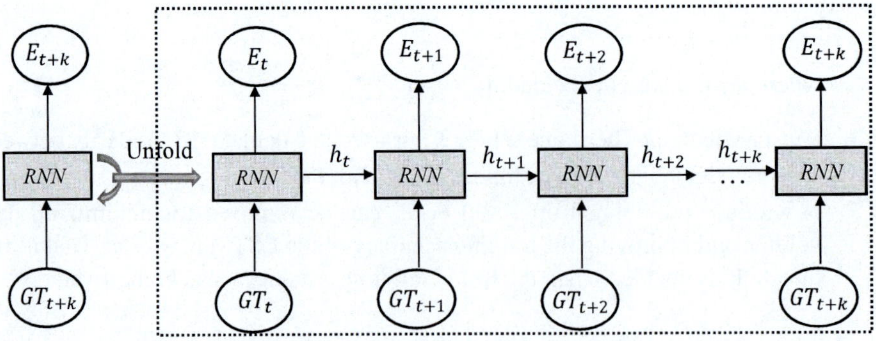

Fig. 5 The general structure of a RNNs

where σ is a sigmoid activation function; t is the time step; T'' is an input vector at time t; h_{t-1} and h_t are hidden layer vectors at times t-1 and t, respectively; W_h and U_h are weight matrices for T'' and h_{t-1}, respectively; b_h is a bias matrix; E_t is an output vector; W_E and b_E are weight and bias vectors, respectively, to calculate E_t.

However, since the hidden state can be extremely small or high with long steps of multiplication, prediction results of the RNNs are usually crippled by a vanishing or exploding gradients problem [27]. To avoid extreme values through the multiplication process, a gating mechanism is required to control which information to be kept (i.e., memorizing) and which information to be removed (i.e., forgetting). A LSTM networks is therefore developed to improve the RNNs design. In a LSTM networks, there are three gates, i.e., input gate, forget gate and output gate, to determine how much information can be fed forward to the next cell [28].

In the forget gate f_t, some information is removed by using a sigmoid activation function:

$$f_t = \sigma(W_{GTf} \cdot GT_t + W_{hf} \cdot h_{t-1} + b_f) \tag{11}$$

where σ is a sigmoid activation function; t is the time step; h_{t-1} is the short-term memory; W_{GTf} and W_{hf} are the weight matrices for GT_t and h_{t-1}, respectively; b_f is the bias in the forget gate.

In the input gate i_t, the information to input is selected by utilising sigmoid and tanh:

$$i_t = \sigma(W_{GTt} \cdot GT_t + W_{ht} \cdot h_{t-1} + b_t) \tag{12}$$

$$\widetilde{C}_t = tanh(W_{GTc} \cdot GT_t + W_{hc} \cdot h_{t-1} + b_c) \tag{13}$$

where tan is the tanh; W_{GTt}, W_{ht}, W_{GTc} and W_{hc} are the weight matrices; b_t and b_c are the bias; $\widetilde{C}_t$ is the output of the standard recurrent layer without the LSTM.

Based on C_{t-1}, $\widetilde{C}_t$, f_t and i_t, the long-term memory C_t can be calculated:

$$C_t = f_t \otimes C_{t-1} \oplus i_t \otimes \widetilde{C}_t \tag{14}$$

where $\otimes$ and $\oplus$ are element-wise multiplication and addition, respectively.

In the output gate O_t, the short-term memory h_t can be calculated below:

$$O_t = \sigma(W_{GTo} \cdot GT_t + W_{ho} \cdot h_{t-1} + b_o) \tag{15}$$

$$h_t = O_t \otimes tanh(C_t) \tag{16}$$

Table 4 Input and output of the LSTM networks

Input vector (GT_t)	Output vector (E_t)
Point 1 in normalized temperature data NT_1	Thermal error data on the X axis E_{t_X}
Point 2 in normalized temperature data NT_2	Thermal error data on the Y axis E_{t_Y}
……	Thermal error data on the Z axis E_{t_Z}
Point n' in normalized temperature data $NT_{n'}$	

$$E_t = h_t \tag{17}$$

where W_{GTo} and W_{ho} are weight matrices; b_o is the bias; E_t is the predicted thermal deformation at time index t.

In this research, the designed LSTM networks has one layer. Its performance is benchmarked with different LSTM layers and the relevant results will be shown in Sect. 3 Chap. 4. Table 4 shows the input and output data of the designed LSTM networks. The normalized temperature data $NT_1, NT_2 \ldots, NT_n$, which are grouped into GT_t, are fed into the LSTM networks as input. The output is the thermal error E_t.

The LSTM networks will be trained and needs to be re-trained (calibration) after a period using historical data. Mean Square Error (MSE), which calculation is based on the difference between predicted thermal errors and practically measured thermal errors, is defined below:

$$MSE = \frac{\sqrt{(E_{t\_X} - P_{t\_X})^2 + (E_{t_Y} - P_{t\_Y})^2 + (E_{t_Z} - P_{t\_Z})^2}}{3} \tag{18}$$

where P_{t_X}, P_{t_Y} and P_{t_Z} are the practically measured thermal errors; E_{t_X}, E_{t_Y} and E_{t_Z} are the predicted thermal errors.

The LSTM networks will be re-trained by using an improved gradient descent optimizer (we improved the gradient descent optimizer in Step 4 with a decaying learning rate to improve the convergence speed). The optimization process is explained as follows:

1. Define the weight W_{LSTM} and bias b_{LSTM} of the LSTM networks; initialize the W_{LSTM} and b_{LSTM} with random values;
2. Define the maximum epoch time T;
3. Optimize the value of W_{LSTM}, b_{LSTM} by minimising the loss function with gradient descent:

$$W_{LSTM}{}^{t+1} = W_{LSTM}{}^{t} - \varepsilon \cdot \left(\frac{\partial(MSE)}{\partial(W_{LSTM})}\right) \quad (19)$$

$$b_{LSTM}{}^{t+1} = b_{LSTM}{}^{t} - \varepsilon \cdot \left(\frac{\partial(MSE)}{\partial(b_{LSTM})}\right) \quad (20)$$

where t is the current epoch time; ε is the learning rate; ∂ is the gradient descent. Traditionally, the learning rate is a constant value.

4. To improve the optimization process, the learning rate ε can be designed to decay over training epochs to approach optimal parameters:

$$\varepsilon = \varepsilon_0/(1 + k \cdot t) \quad (21)$$

where k is the hyper-parameter.

5. Repeat the above Steps 3 and 4 until reaching the maximum epoch time.

3.4 Fog-Cloud Architecture Design

A cloud structure has been widely used to process computationally intensive tasks benefiting from its centralized storages and strong computational capacities on the cloud server side. However, the performance of the cloud solution could be severely compromised by the limited communication bandwidth of the industrial Internet in manufacturing companies. Furthermore, data security in cloud is also a key concern for industrial users. To address the issues, a fog-cloud architecture, which consists of a terminal layer, a fog layer and a cloud layer, is deployed to support the thermal error prediction system. The architecture is illustrated in Fig. 6. Some details are explained below:

- Terminal layer: The terminal layer is integrated with the physical heavy-duty machine. In the layer, sensors are mounted on the machine (the details are explained in Sects. 1 and 2). The machine undergoes machining processes, and thermal sensors and laser displacement sensors are installed on the machine to continuously collect data. The optimal placement of thermal sensors will be finally guided by the result of FEA, which shows heat distributions of the machine over the entire machining process. The data are transmitted to the fog layer through an Internet router for further processing.
- Fog layer: Cost-effective edge devices are deployed on the fog layer to provide certain computational capability for local processing on the transmitted data. Local processors include a pre-processor *i*GRA and a trained LSTM networks. Based on the correlation among thermal data, similar data are removed by *i*GRA to minimize data collinearity and optimize the overall computing efficiency. The trained LSTM networks is used for thermal error prediction. Considering dynamics in thermal

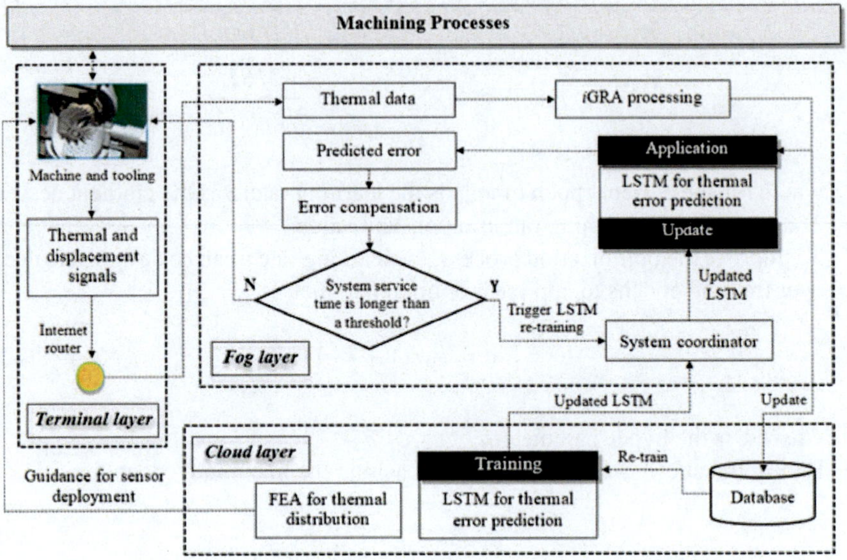

Fig. 6 The fog-cloud architecture enabled thermal error prediction and compensation

changes throughout machining processes, after a period of system service, a predefined threshold, the LSTM networks will be re-trained on the cloud layer based on updated time-series sensor data.

- Cloud layer: The cloud layer hosts a cloud server providing intensive computational power and storage spaces, so that all the processing that requires high computational resources and data storage will be done on the cloud layer. FEA is processed on the cloud layer to identify the heat distribution over the entire structure of CNC machine. The results will be sent back to the terminal layer to provide an instruction to optimize thermal sensor deployment. The LSTM networks are re-trained on the cloud layer when necessary. A system coordinator, which connects the fog layer and the cloud layer, triggers the training process of the LSTM networks with an embedded improved gradient descent optimizer.

To check the improvement of machining accuracy of the heavy-duty machine by using the proposed system, thermal errors without using the system on the X, Y and Z axes are measured as$E_{t_without} = [E_{t_{X_without}}, E_{t_{Y_without}}, E_{t_{Z_without}}]$, and the thermal errors with the proposed system on the X, Y and Z axes are measured as$E_{t_with} = [E_{t_{X_with}}, E_{t_{Y_with}}, E_{t_{Z_with}}]$. The improvement of the machining accuracy of the proposed system can be calculated:

$$Improvement = \frac{\sum_{i=1}^{3} (E_{t_{without_i}} - E_{t\_with_i})^2}{3} \quad (22)$$

Table 5 Detailed specifications for deployed sensors and data processing system

Terminal layer	Sensor type	Sensitivity (pm/°C)	Accuracy (°C)	Resolution (°C)
	Fiber Bragg grating (FBG)	38.1	≤ 0.1	≤ 0.05
Fog layer	Processor	Speed	Memory size	RAM
	FPGA: Cyclone II EP2C5T144C8N	260 MHz	119,808 bit	1.1 MB
Cloud layer	Communications	Processor	Memory	Storage
	2.4 GHz/5 GHz IEEE 802.11	4.2Ghz	32 GB	1 TB

In this research, specifications to implement the fog-cloud architecture are presented in Table 5. The method to quantify sensitivity and accuracy are as follows:

$$Sensitivity = \frac{Reflectionwavelengthshift}{\Delta T} \tag{23}$$

$$Accuracy = T_{true} - T_{sensor} \tag{24}$$

where ΔT is the change in temperature, T_{true} and T_{sensor} are the true temperature and the measured temperature by sensor, respectively. Resolution is the smallest change in temperature that the sensor can detect.

On the terminal layer, thermal sensors of Fibre Bragg gratings (FBGs) are deployed. The sensors are of small size, flexibility, immunity to electro-magnetic interference, etc., and they also have good accuracy ($\le$0.1 °C). On the fog layer, the processor FPGA CycloneII EP2C5T144C8N, is deployed. FPGA exhibits good performances in executing the LSTM networks locally as it is of robust flexibility, reconfigurability and efficient parallel computing, while it is more cost effective in comparison with those processors deployed on the cloud layer. Training on the LSTM networks and FEA computations are computationally intensive, so that the LSTM training/re-training tasks and FEA are assigned to the cloud layer to leverage its better computational speed and larger memory.

4 Case Studies and Experimental Analysis

4.1 Sensor Deployment

As aforementioned, the heavy-duty machine ZK5540 is used as an example to explain and verify the developed methodologies. After the FEA analysis for optimization, there are 128 thermal sensors installed on ZK5540. The positions of the groups of sensors are marked in Fig. 7a. More details are:

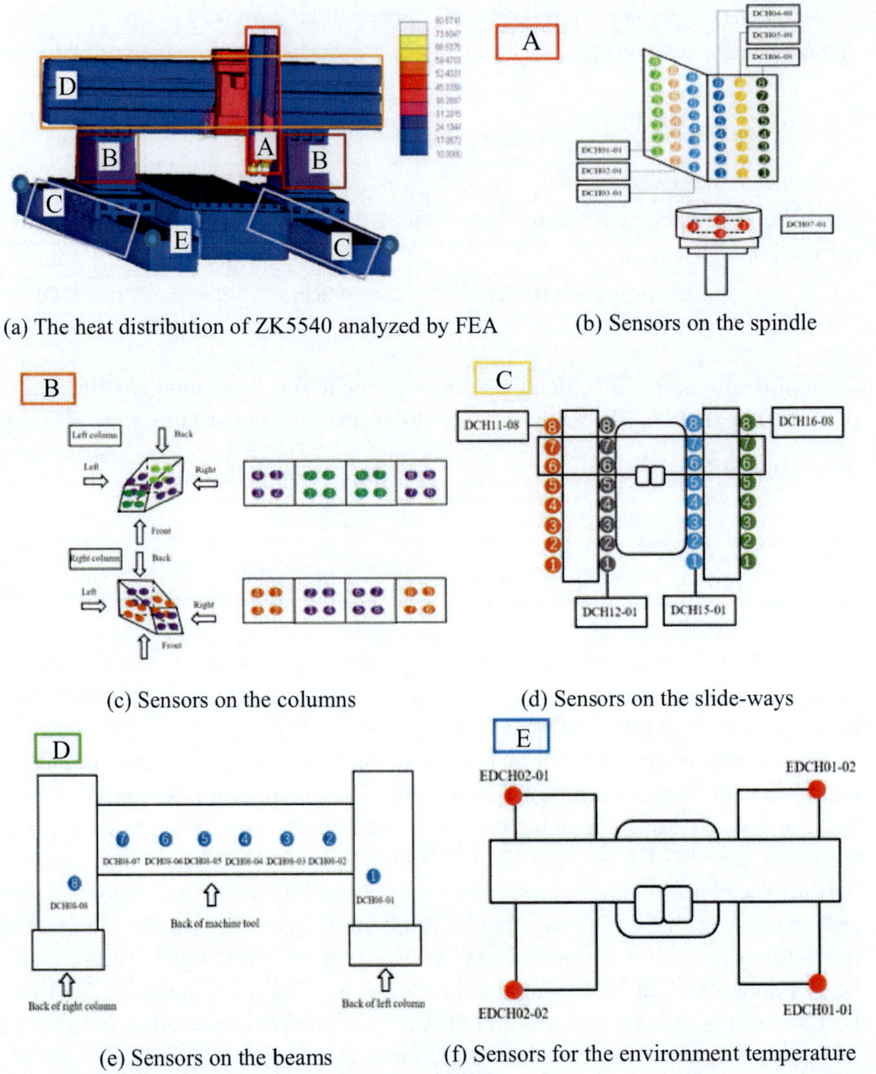

Fig. 7 Thermal sensor placement

- 52 sensors are installed on the spindle (as shown in Fig. 7b);
- 32 sensors are installed on the two columns, and each with 16 (as shown in Fig. 7c);
- 32 sensors are installed on the two slide-ways, and each with 16 (as shown in Fig. 7d);
- 8 sensors are installed on the beam (as shown in Fig. 7e);
- 4 sensors are installed around machine tool to measure environmental temperature, and each side with one (as shown in Fig. 7f).

Laser displacement sensors are installed on the machining bed to measure the thermal deformation on the X, Y and Z directions.

All the sensor data points are presented in sequences $\boldsymbol{T} = [\sum_{j=1}^{m} t_{1,j}, \ldots, \sum_{j=1}^{m} t_{128,j}]$, where j is the time index and m is the maximum time index. Among them,

- $\sum_{j=1}^{m} t_{1,j}$ to $\sum_{j=1}^{m} t_{52,j}$ are the sensor data on the spindle,
- $\sum_{j=1}^{m} t_{53,j}$ to $\sum_{j=1}^{m} t_{84,j}$ are the sensor data on the columns,
- $\sum_{j=1}^{m} t_{85,j}$ to $\sum_{j=1}^{m} t_{116,j}$ are the sensor data on the slide-ways,
- $\sum_{j=1}^{m} t_{117,j}$ to $\sum_{j=1}^{m} t_{124,j}$ are the sensor data on the beam,
- $\sum_{j=1}^{m} t_{125,j}$ to $\sum_{j=1}^{m} t_{128,j}$ are the sensor data around the four corners of the machine.

Experiments were conducted over three months. 9.28 GB temperature data were collected. Each data node generates 0.0725 GB data over the period.

The counting sequence of sensors on each area is in the directions from left to right and from upward to downward. For example, the presentation of sensor data on the spindle is illustrated in Fig. 8. Figure 9 shows that the laser displacement sensors measure the thermal deformation of the spindle during machining. The data are collected and transmitted to the fog layer through the Internet router for local processing. Data showing temperature and thermal errors on the spindle over time are displayed in Fig. 10. Some observations can be made as follows: (1) the thermal errors are sensitive to the temperature changes; (2) the changing patterns of temperatures and thermal errors over time are closely correlated; (3) a number of temperature

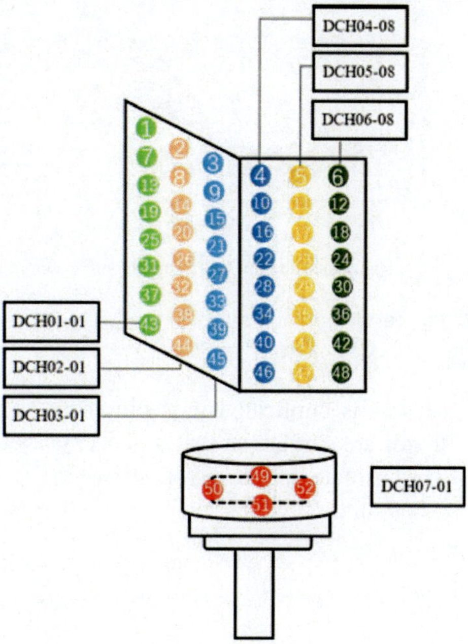

Fig. 8 Thermal sensor deployment on the spindle

Fig. 9 Deployment of laser displacement sensors for thermal deformation measurement

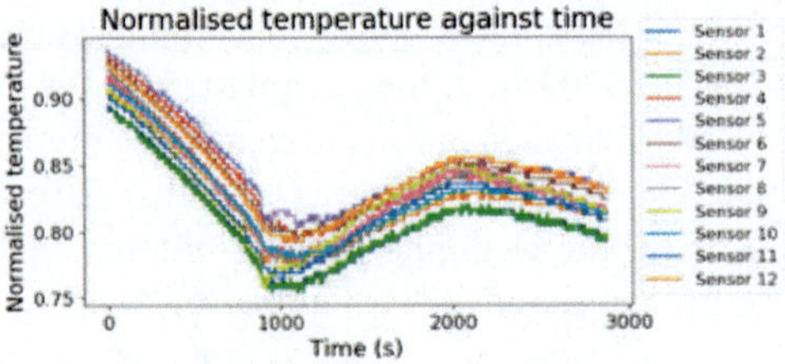

(c) Normalized temperature data (input): example 2

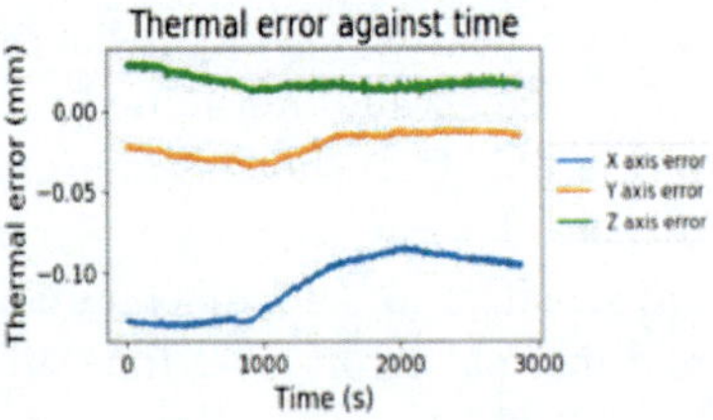

(d) Thermal error data (output): example 2

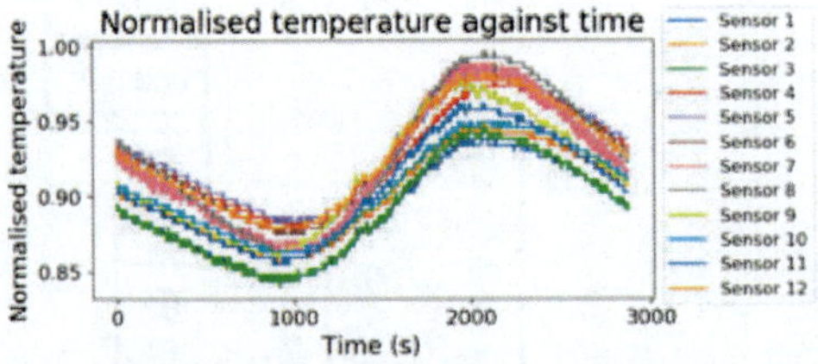

(a) Normalized temperature data (input): example 1

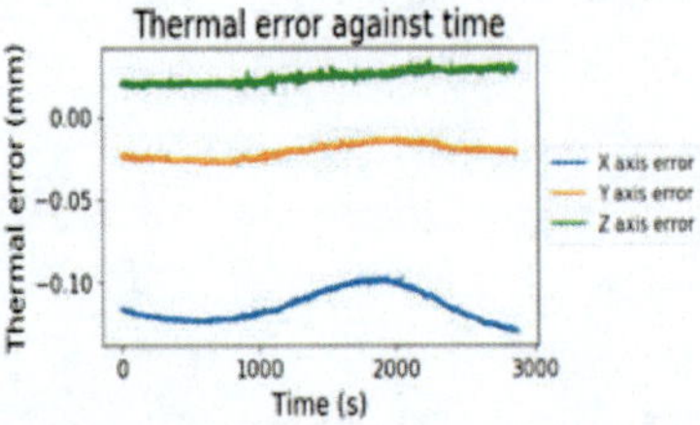

(b) Thermal error data (output): example 1

Fig. 10 Temperature data and thermal errors

sensors is duplicated in deployment as the measured temperatures and changing trends are similar, so that it is necessary to reduce the number of sensors by *i*GRA. The absolute thermal errors shown in Fig. 10 are in the range between 0.134 mm and 0.009 mm.

4.2 Analysis on iGRA

For the collected data from the 128 thermal sensors on the terminal layer, they are sent to the fog layer for pre-processing by *i*GRA. Figure 11 illustrates the total grey relational grade among 1–12 sensors on the spindle according to Eqs. (4)–(6). Based on statistics, there are 12 pairs, 28 pairs, 53 pairs and 88 pairs of sensor data for more than 0.9, 0.8, 0.7 and 0.6 in terms of total grey relational grade, respectively. It is paramount to determine the value of an appropriate threshold in order to ensure optimal data transmission and the training accuracy of the LSTM networks. In this case, the training accuracy of the LSTM networks is set as priory. Therefore, the threshold that achieves the highest training accuracy will be selected.

For this case, there are 8,128 grey relational grades calculated in total according to Eq. (6). To determine the best threshold for sensor selection, different values of threshold were compared in terms of removed sensor data nodes, saved transmitted data and final training accuracy according to Eq. (8). In this application, the final training accuracy is more important, so that $w_1 = 0.1$ and $w_2 = 0.9$ according to Eq. (8). The results are shown in Table 6. The details of training accuracy with the designed LSTM networks are shown in Fig. 12, where each dataset is utilised for

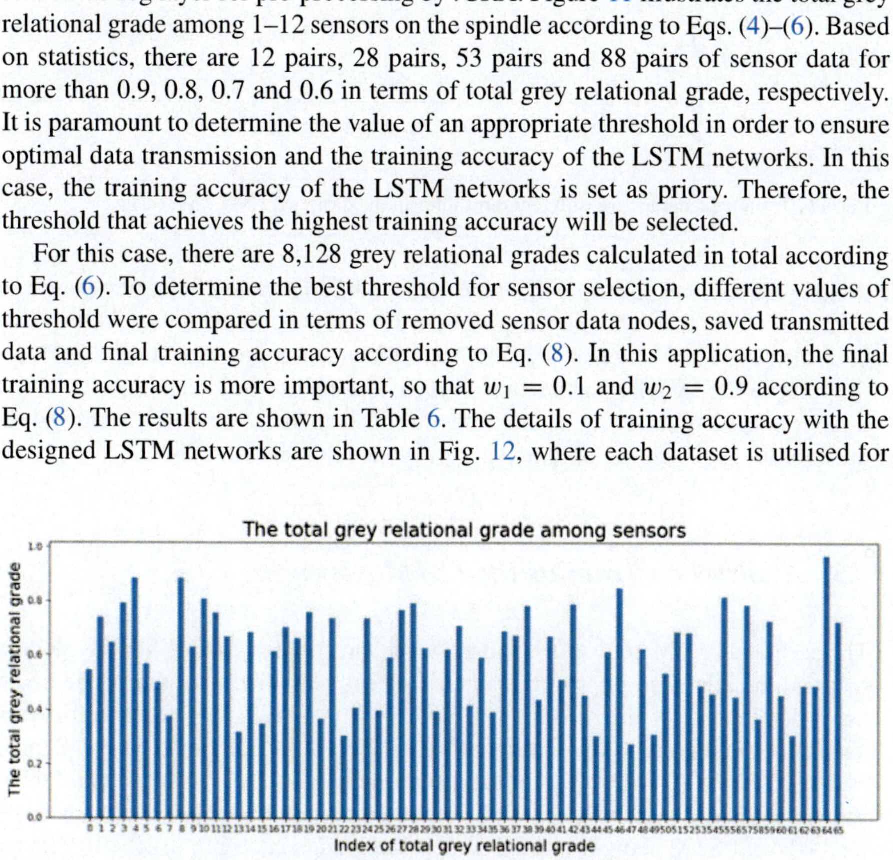

Fig. 11 The total grey relational grade among 1–12 sensors

Table 6 Results of iGRA using various threshold settings

Threshold	Removed sensors	Saved data transmission (GB)	Final training accuracy (%)	Fitness
0.9	12	0.87	89	0.50
0.85	16	1.16	84	0.01
0.8	28	2.03	92	0.82
0.75	40	2.90	87	0.33
0.7	53	3.84	91	0.75
0.65	69	5.00	93	0.97
0.6	88	6.38	91	0.80

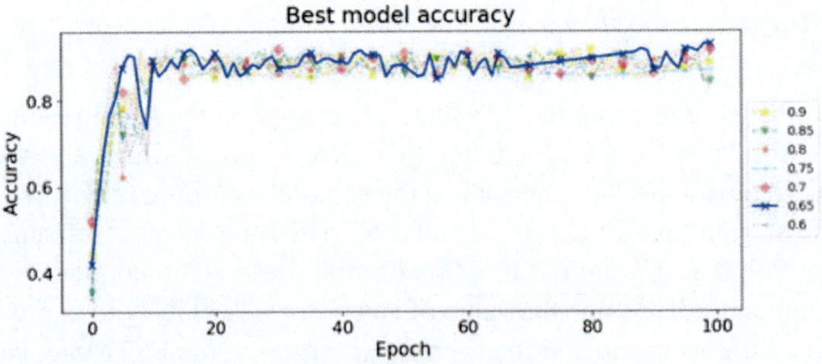

Fig. 12 Training accuracy for different thresholds in the designed LSTM networks

training for 10 times to ensure the robustness of the approach (10 times was the most used times for training in many applications [29]. It indicates that when threshold = 0.65, the fitness is the highest (0.97). That is, the final training accuracy is 93.0% and the volume of 5.00 GB can be reduced when data are transmitted to the cloud layer. It represents 52.63% saving in total data transmission. It can be clearly observed that the threshold can achieve optimum accuracy quickly with robustness.

4.3 Analysis on Training the LSTM Networks

The designed LSTM networks is trained when threshold = 0.65. When the model is required to update, an improved gradient descent is applied to optimize the LSTM networks. To justify the selection of the LSTM design in this research, different algorithms, including RNNs, Artificial Neural Network (ANN), CNNs and Support Vector Machine (SVM), were utilized for benchmarking. Figure 13 and Table 7 show the training details. It shows that both the best and average training accuracy for a 1-layer LSTM networks with the improved gradient decent can achieve 93.1 and 88.9%, which are the highest among the comparative algorithms. Compared with the 1-layer LSTM networks, a 2-layer LSTM networks is 25.46% lower in the average accuracy and a 3-layer LSTM is 2.90% lower in the average accuracy.

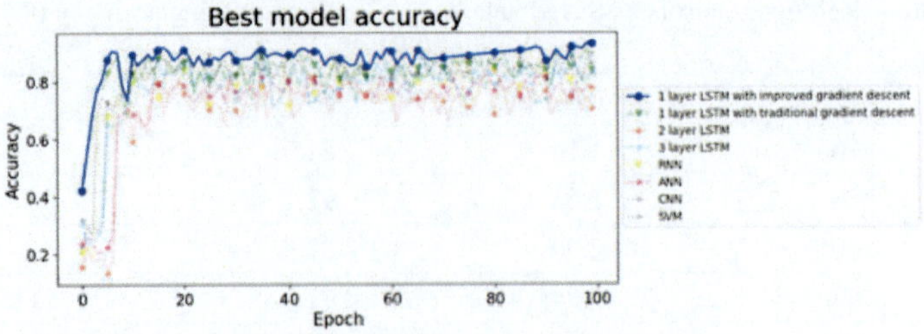

Fig. 13 Training accuracy of different algorithms

Table 7 Benchmarking for different algorithms

Algorithms	Best accuracy (%)	Worst accuracy (%)	Average accuracy (%)
1-layer LSTM with an improved gradient descent	93.1	85.2	88.9
1-layer LSTM with a traditional gradient descent	85.0	81.9	83.9
2-layer LSTM	69.4	67.1	68.8
3-layer LSTM	90.4	87.6	89.5
RNN	85.3	83.1	84.3
ANN	77.5	71.2	76.7
CNN	89.6	81.0	84.3
SVM	85.9	82.0	83.9

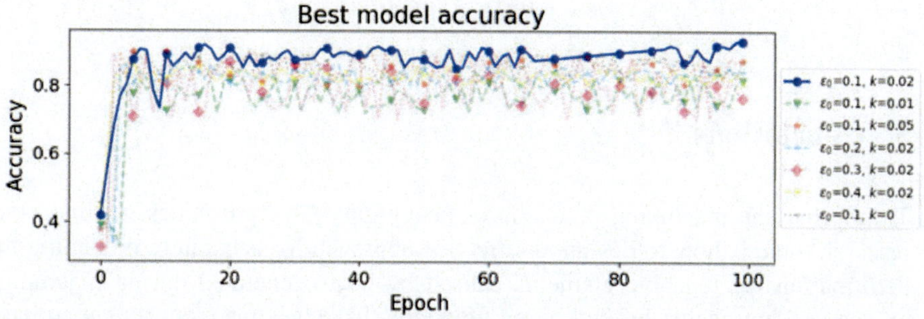

Fig. 14 Training accuracy of different training parameters

Experiments were also carried out to select optimal parameters in Eqs. (19)–(21). Figure 14 and Table 8 illustrate the training details. It shows that, when training parameters are ε_0 = 0.1 and k = 0.02, the best and average accuracy can achieve 93.1% and 88.9%, which are the highest among the results in different parameter settings. It can be seen that when ε_0 = 0.1 and k = 0.01, the best accuracy is the lowest (73.4%). The proposed parameter has 19.70% higher best training accuracy than that of the lowest best accuracy. Based on the proposed method, they system is deployed in production process, the thermal errors before and after deploying the system ($E_{t\_without}$ and E_{t_with}) have been measured. According to calculation Eq. (16), 46.53% improvement in machining accuracy was achieved.

Table 8 Benchmark for different training parameters

Parameters	Best accuracy (%)	Worst accuracy (%)	Average accuracy (%)
$\varepsilon\_0 = 0.1$, $k = 0.02$	93.1	85.2	88.9
$\varepsilon\_0 = 0.1$, $k = 0.01$	73.4	68.5	72.0
$\varepsilon\_0 = 0.1$, $k = 0.05$	84.2	75.0	80.7
$\varepsilon\_0 = 0.2$, $k = 0.02$	83.8	69.1	77.5
$\varepsilon\_0 = 0.3$, $k = 0.02$	80.4	73.4	78.0
$\varepsilon\_0 = 0.4$, $k = 0.02$	85.1	72.6	80.0
$\varepsilon\_0 = 0.1$, $k = 0$	86.6	77.8	83.4

5 Conclusions

To improve the machining precision of heavy-duty CNC machines, an important research topic is how to design an effective approach for accurately predicting the deformations of machine elements caused by heats generated during machining processes. Approaches of data-based modelling have become increasingly adopted but they exhibit low efficiency in processing large-volume thermal data. To tackle the issue, this chapter presents a new system enabled by a LSTM networks and fog-cloud architecture for thermal error prediction of heavy-duty machines. The system has the following characteristics: (1) The system uses physics-based modelling to optimise data-based modelling in the aspect of deciding the optimal number/locations of deployed thermal sensors and minimised measured data; (2) The LSTM networks is integrated with an effective data pre-processor to improve thermal error prediction in terms of prediction accuracy and data processing efficiency; (3) The fog-cloud architecture can improve the system efficiency of data transmission and system responsiveness to compensate thermal errors on the fly. The developed system was validated using an industrial heavy-duty CNC machine. Based on industrial case studies, the volume of transmitted data was minimised by 52.63% and the machining accuracy was improved by 46.53%, in comparison with the processes without using the developed system. It proves that the designed system has a great potential in supporting real-world industrial applications.

In future research, the following aspects will be further investigated to improve the system:

- Computation resource optimisation between both fog layer and cloud layer will be further researched to achieve optimum computation resource distribution among the layers;

- To enhance the machining error prediction, different data source (e.g., vibration, contact force, etc.) could be fused for analysis;
- This research investigates machining inaccuracy from the aspect of thermal errors. There are more factors that influence machining accuracy, such as machining condition, machining shapes, schedules, etc. These will be analysed in the future.

References

1. Huang J, Zhou ZD, Liu M, Zhang E, Chen M, Pham DT, Ji C (2015) Real-time measurement of temperature field in heavy-duty machine tools using fiber bragg grating sensors and analysis of thermal shift errors. Mechatronics 31:16–21
2. Bryan J (1990) International status of thermal error research. CIRP Ann 39(2):645–656
3. Li Y, Zhao W, Lan S, Ni J, Wu W, Lu B (2015) A review on spindle thermal error compensation in machine tools. Int J Mac Tools Manufact 95:20–38
4. Li F, Li T, Jiang Y, Wang H, Ehmann K (2019) Explicit error modeling of dynamic thermal errors of heavy machine tool frames caused by ambient temperature fluctuations. J Manufact Process 48:320–338
5. Miao E, Liu Y, Liu H, Gao Z, Li W (2015) Study on the effects of changes in temperature-sensitive points on thermal error compensation model for CNC machine tool. Int J Mach Tools Manuf 97:50–59
6. Cheng Q, Qi Z, Zhang G, Zhao Y, Sun B, Gu P (2015) Robust modelling and prediction of thermally induced positional error based on grey rough set theory and neural networks. Int J Adv Manufact Technol 83(5–8):753–764
7. Abdulshahed A, Longstaff A, Fletcher S, Myers A (2015) Thermal error modelling of machine tools based on ANFIS with fuzzy c-means clustering using a thermal imaging camera. Appl Math Modell 41:130–142
8. Liu Q, Yan J, Pham DT, Zhou ZD, Xu W, Wei Q, Ji C (2015) Identification and optimal selection of temperature-sensitive measuring points of thermal error compensation on a heavy-duty machine tool. Int J Adv Manufact Technol 85:345–353
9. Ma C, Zhao L, Mei X, Shi H, Yang J (2016) Thermal error compensation of high-speed spindle system based on a modified BP neural network. Int J Adv Manufact Technol 89:3071–3085
10. Li F, Li T, Wang H, Jiang Y (2017) A temperature sensor clustering method for thermal error modelling of heavy milling machine tools. Appl Scie 7(1):82
11. Li B, Tian X, Zhang M (2019) Thermal error modeling of machine tool spindle based on the improved algorithm optimized BP neural network. Int J Adv Manufact Technol 105:1497–1505
12. Fujishima M, Narimatsu K, Irinoa N, Mori M, Ibaraki S (2019) Adaptive thermal displacement compensation method based on deep learning. CIRP J Manufact Sci Technol 25:22–25
13. Yao X, Hu T, Yin G, Cheng C (2020) Thermal error modeling and prediction analysis based on OM algorithm for machine tool's spindle. Int J Adv Manufact Technol 106:3345–3356
14. Qu Z, Zheng S, Wang X, Song X, Li B, Song X (2018) Converged Recommendation System Based on RNN and BP Neural Networks. 2018 IEEE International Conference on Big Data and Smart Computing
15. Zhang J, Wang P, Yan R, Gao R (2018) Long short-term memory for machine remaining life prediction. J Manufact Syst 48:78–86
16. Li F, Xiang W, Wang J, Zhou X, Tang B (2018) Quantum weighted long short-term memory neural network and its application in state degradation trend prediction of rotating machinery. Neural Net 106:237–248
17. Yang J, Guo Y, Zhao W (2019) Long short-term memory neural network based fault detection and isolation for electro-mechanical actuators. Neurocomputing 360:85–96

18. An Q, Tao Z, Xu X, El Mansori M, Chen M (2020) A data-driven model for milling tool remaining useful life prediction with convolutional and stacked LSTM network. Measurement 154:107461
19. Xia T, Song Y, Zheng Y, Pan E, Xi L (2020) An ensemble framework based on convolutional bi-directional LSTM with multiple time windows for remaining useful life estimation. Comput Ind 115:103182
20. Liu C, Zhu L (2020) A two-stage approach for predicting the remaining useful life of tools using bidirectional long short-term memory. Measurement 164:108029
21. Hu P, Dhelim S, Ning H, Qiu T (2017) Survey on fog computing: architecture, key technologies, applications and open issues. J Netw Comput Appl 98:27–42
22. Bellavista P, Berrocal J, Corradi A, Das S, Foschini L, Zanni A (2019) A survey on fog computing for the Internet of Things. Pervas Mobile Comput 52:71–99
23. Liang YC, Li WD, Lu X, Wang S (2019) Fog computing and convolutional neural network enabled prognosis for machining process optimization. J Manufact Syst 52:32–42.
24. Wu D, Liu S, Zhang L, Terpenny J, Gao R, Kurfess T, Guzzo J (2018) A fog computing-based framework for process monitoring and prognosis in cyber-manufacturing. J Manufact Syst 43:25–34
25. Vlachas P, Pathak J, Hunt B, Sapsis T, Girvan M, Ott E, Koumoutsakos P (2020) Backpropagation algorithms and reservoir computing in recurrent neural networks for the forecasting of complex spatiotemporal dynamics. Neural Netw 126:191–217
26. Cinar Y, Mirisaee H, Goswami P, Gaussier E, Aït-Bachir A (2018) Period-aware content attention RNNs for time series forecasting with missing values. Neurocomputing 312:177–186
27. Mohanty S, Lydia E, Elhoseny M, Al OM, Shankar K (2020) Deep learning with LSTM based distributed data mining model for energy efficient wireless sensor networks. Physi Commun 40:101097
28. Ellis M, Chinde V (2020) An encoder–decoder LSTM-based EMPC framework applied to a building HVAC system. Chem Eng Res Des 160:508–520
29. Kalamatianos R, Kermanidis K, Karydis I, Avlonitis M (2018) Treating stochasticity of olive-fruit fly's outbreaks via machine learning algorithms. Neurocomputing 280:135–146

Deep Transfer Learning Enabled Estimation of Health State of Cutting Tools

M. Marei, S. El Zaataria, and W. D. Li

1 Introduction

In manufacturing industries, unplanned downtime is known to negatively impact profitability, and will be a barrier to implementing lean and zero-defect manufacturing. Also, operational safety could be compromised through unexpected failures, particularly when human operators are involved [1]. To tackle the challenge, predictive maintenance based on Prognostics and Health Management (PHM) has been developed to predict the failure points of working components (such as bearings and cutting tools) [2–6]. Based on that, a component in a manufacturing system can be replaced just before it fails. Thus, component lifetime can be maximized, system downtime can be minimized, and therefore optimal productivity and production quality can be achieved. In Computerized Numerical Control (CNC) machining processes, cutting tool wear leads to various manufacturing problems, ranging from stoppage downtime for redressing and tool replacement, to scraps and reworks of machined components due to degradation in surface quality [7]. Therefore, accurate prediction of the Remaining Useful Life (RUL) for a cutting tool is essential to mitigate such failures [8].

In the aspect, physics-based approaches on empirical models have been developed, such as the Taylor, Extended Taylor, Colding, and Coromant Turning model (a more detailed review can be found in [9]). However, these approaches are sensitive to variations in machining parameters (e.g., cutting speed, feed rate, cutting depth, cutting tool properties such as the number of teeth), which vary depending

M. Marei · S. E. Zaataria · W. D. Li (✉)
Faculty of Engineering, Environment and Computing, Coventry University, Coventry, UK
e-mail: weidong.li@coventry.ac.uk

W. D. Li
School of Logistics Engineering, Wuhan University of Technology, Wuhan, China

W. Li et al. (eds.), *Data Driven Smart Manufacturing Technologies and Applications*, Springer Series in Advanced Manufacturing,
https://doi.org/10.1007/978-3-030-66849-5_7

on component materials and machining set-up. Moreover, profound expert knowledge of machining processes is also expected to conduct effective and accurate RUL prediction. In contrast to physics-based approaches, data-driven approaches have been developed to leverage historical and real-time data to support decision-making. Deep learning algorithms (e.g., Convolutional Neural Networks (CNNs)) have been explored to facilitate data-driven approaches (a related review can refer to [10]). For instance, to attain a wide scope of image features from a variety of applications, CNNs models, such as ImageNet [11] (and the subsequent ImageNet Large Scale Visual Recognition Challenge (ILSVRC) [12]), CIFAR-10 and CIFAR-100 [13], can be trained on millions of images of natural and synthetic objects. These approaches excel at extracting discriminative features and learning hidden relationships between problem parameters (i.e., feature learning), opposed to feature engineering approaches where human experts specify the features to learn [14]. However, the accuracy and reliability of deep learning enabled data-driven approaches may be reduced significantly when data are scarce or insufficiently descriptive of the problem. To address the issue, in recent years, transfer learning enabled approaches have been developed to improve pre-trained deep learning models to perform prediction on new problems with limited available data [15, 16]. The transfer learning strategy is to retain and reuse domain-invariant features learn from a task in one domain (called the source domain), to enhance the performance of another task in its corresponding domain (called the target domain). On the other hand, though transfer learning has shown its great potentials in different applications, there are still limited research works reported on manufacturing applications, especially for estimating the RUL of cutting tools.

In this chapter, it is aimed to develop a transfer learning enabled CNNs approach for RUL estimation of cutting tools and further establish PHM based on analyzing the images of healthy and worn tools. The problem of limited data available during machining processes is tackled by integrating transfer learning into CNNs. The main characteristics of this research are summarized as follows:

1. Developing a transfer learning enabled CNNs approach with gradient descent optimization algorithms based on normalized Maximum Mean Discrepancy (MMD) and Mean Square Error (MSE) for algorithm training with learnable weights;
2. Benchmarking the performance of the approach through evaluating several CNNs architectures and transfer learning for tool RUL prediction and PHM;
3. Providing recommendations for training techniques and data augmentation based on benchmarking and analysis results.

2 Related Works

A brief review on prior works in deep learning for PHM applications is presented here with a focus on implementations in the manufacturing domain. Relevant literature aligning with CNNs and transfer learning are also highlighted.

2.1 Deep Learning for PHM

In recent years, there have been increasing reviews on investigating the development of deep learning approaches for PHM applications [10]. Most deep learning approaches for PHM exploit an inflow of continuous real-time data, such as vibration signals [5], acoustic emission sensor data [8], force measurements [7, 17], temperature [8], power/current, etc. Alternatively, other approaches were developed and tested with an existing RUL dataset at the validation stage (some examples of these datasets are mentioned in [9]). The C-MAPSS (aero-engine health degradation simulation) tool [17] was used to create datasets extensively studied in prior works on PHM, with varying research perspectives [18]. In addition, the PHM society holds an annual challenge for PHM techniques based on data that it provides, in which the 2010 dataset focuses on high-speed CNC machining [19]. While numerous architectures of deep learning were implemented for PHM, several primary models can be summarized into the following types:

- CNNs and their variants: these approaches use a series of shallow or deep convolutional filters that extract spatial features from images or image-like data. These approaches are particularly efficient at learning meaningful representations of the data from individual and grouped clusters of filters across several depth channels [20]. Deeper CNNs layers are typically capable of extracting distinct visual features like parts of the general shape of the image, whereas shallower layers typically extract image intensity or color variation across a limited image window [21]. While predominantly used for failure classification in PHM [22], several researchers successfully used CNNs for regression-related feature extraction for RUL prediction [23]. A few approaches were developed to perform both classification and regression [23]. However, few approaches utilized images as input for predicting the health state of cutting tools.
- Recurrent Neural Networks (RNNs and their variants: these models learn from data sequences with explicit time dependencies between input samples (i.e., sequence prediction problems), due to series of gates within the architecture which retain the outputs from previous inputs. The output from one neuron is fed forward into the adjacent neuron, so that the hidden representation or the output of the neuron is influenced by past inputs [11]. Long Short-Term Memory (LSTM) networks [24] are similar to RNNs but can retain memory over a longer time horizon due to a more complex gate structure that preserves long-term dependencies. A combination of gates with different activation functions determine what portion of the cell state to retain or transform for input into the subsequent layer. LSTM and their variants have achieved widespread success at time series-based prediction problems, therefore are especially popular for PHM applications [21].
- Auto-Encoders (AEs) and their variants: AEs are feedforward neural networks comprising an encoder and a decoder. The encoder is trained to find the hidden representation of the input data via a nonlinear mapping of the weights and biases. The decoder attempts to find an output to the inverse function of the hidden representation to the data, i.e., a nonlinear mapping between the encoder function

and the hidden representation. To avoid human-assisted feature identification for Tool Condition Monitoring (TCM), Shi et al. used a novel Feature Multi-Space Sparse Auto-Encoder (FMSSAE) to classify four different wear states from TCM data in the time domain (TD), frequency domain (FD) and Wavelet Domain (WD) [25]. Their methodology reported to achieve 96% accuracy in classification.

- Hybrid approaches: which combine and adapt several architecture designs to conduct feature extraction and time-based prediction, e.g., Zhao et al. used a Convolutional Bi-directional Long Short-Term Memory (CBLSTM) to extract local and informative sequence features, followed by a bi-directional LSTM to identify long-term time dependencies, leading to linear regression for target value prediction [6]. The approach was demonstrated to predict tool wear based on raw sensory data.

2.2 *Transfer Learning Enabled Deep Learning*

In the reviewed literature, an ongoing theme is to develop approaches that are trained or validated on publicly available datasets or those collected in individual experiments. Consequently, the replicability of these studies that leverage closed-source data may be called into question [11]. An additional limitation to the data being generated for PHM studies relates to the classic imbalance problem (whereby healthy data samples are much more prominent than faulty data samples) [11]. Meanwhile, the accuracy and reliability of the approaches are significantly hindered by insufficient manufacturing data: a key limitation for most deep learning approaches is their reliance on large quantities of data, typically in the order of ~1000 samples per class (for classification type problems).

Transfer learning has presented its potential to address the above problems [15, 16, 26]. With transfer learning, knowledge acquired from one domain might be retained and reused in a new domain. In general, the methodologies of transfer learning can be classified into the following four categories according to what and how the knowledge is transferred: (1) Instance based transfer learning—the labelled datasets from the source domain are reused in the target domain; (2) Feature based transfer learning—the features in the source domain are reused in the target domain if the features of the source domain match those in the target domain; (3) Parameter based transfer learning—the setting parameters of a machine learning algorithm in the source domain are re-used in the target domain; (4) Relational knowledge-based transfer learning—the relationship between the data from the source domain and the target domain is established, which is the base for knowledge transfer between the two domains.

In essence, transfer learning models repurpose the weights of deep learning approaches learned in classification tasks, corresponding to features in a similar or different domain (e.g., general-purpose image classification challenge datasets like ILSVRC [8] and Places [27]), to perform predictions for a new task. Such models typically achieved remarkable success, leading to a new research direction in exploring this generalizability by evaluating the classification or regression performance in new tasks (i.e., domain adaptation [27]). In particular, for transfer learning, many approaches work well under a pre-requisite condition: the cross-domain datasets are drawn under the same feature distribution. When the feature distribution changes, most deep learning based approaches need to be re-trained from scratch.

In recent years, various research works have been conducted to integrate transfer learning into deep learning algorithms, i.e., deep transfer learning. For instance, Lu et al. developed a deep transfer learning algorithm based on deep neural network and feature based transfer learning for domain adaptation [28]. In the research, the distribution difference of the features between the datasets from the source domain and target domain was evaluated based on the Maximum Mean Discrepancy (MMD) in a Reproducing Kernel Hilbert Space (RKHS). Weight regularization was carried out by an optimization algorithm to minimize the difference of the MMD for the two domains in implementing knowledge transfer. Xiao et al. designed another deep transfer learning algorithm for motor fault diagnostics [29]. In the algorithm, a feature based transfer learning approach was developed to facilitate knowledge learnt from labelled data under invariant working conditions to the unlabeled data under constantly changing conditions. MMD was incorporated into the training process to impose constraints on the parameters of deep learning to minimize the distribution mismatch of features between two domains. Wen et al. proposed a novel deep transfer learning algorithm for fault diagnosis. A cost function of weighted fault classification loss and MMD was used to fine-tune the model weights [30].

2.3 Images in PHM Applications

Traditionally, images are used within PHM for visual inspection when assessing the condition of the damaged component or machine. However, similar image data could be used as a viable tool to predict (or localize) faults within a machine component, particularly if the frequency of such image measurements is sufficiently large. Despite their recent success in several time–frequency applications for PHM, applications of CNNs to process image data in a PHM context are still not frequent. Most approaches use either 2D representations of time- or frequency-domain data (e.g., engine sensor data in [31], vibration signals in bearings in [5]). Comparatively few examples exist where the failure mode of a machine was classified by a pre-trained CNNs model based on visual data. In particular, Janssens et al. utilized pre-trained versions of the VGG-16 CNNs model [32, 33]. The model was re-trained on thermal video images to classify the failure type, firstly by the image intensity, and secondly based on the degree of imbalance of the rotating machine. Their approach combines a CNNs trained to extract spatial features from thermal images, and another trained to extract temporal information from differenced images to represent the motion due to imbalances. In their first case study, 12 different health conditions were successfully classified with 91.67% accuracy using a deep learning approach, versus conventional approach that classified the 12 health conditions with 55.00% accuracy. Additionally, they implemented the same methodology to perform binary classification on another system, reporting an accuracy of 86.67% with the feature learning vs 80% with the conventional approach. The work provides a promising first step towards intelligent PHM for real-time fault prediction.

The above research predominantly relies on a large number of images to perform RUL prediction or failure classification for diagnosis. While demonstrating widespread success on their individual dataset, these methodologies are ineffective for the cases of limited images available. The methodology in this research leverages transfer learning enabled CNNs for PHM applications, through which the extensibility to other problems within PHM are also demonstrated.

3 Methodology

An overview of the developed methodology is described here, first by introducing the overall workflow and then by detailing its constituent components.

3.1 Problem Definition and Overall Methodology

The overall methodology is shown in Fig. 1. The objective of this research is to predict the RUL of a cutting tool, given the image of the tool as an input and the corresponding normalized tool wear measurement as a prediction target. In other words, the objective is to determine:

$$\widetilde{\boldsymbol{y}}= \boldsymbol{w}^T \boldsymbol{x} + \boldsymbol{b} \tag{1}$$

where $\widetilde{\boldsymbol{y}}$ is the matrix for the predicted regression output, $\boldsymbol{w}$ and $\boldsymbol{b}$ are the trainable weights and biases of a CNNs model respectively, and $\boldsymbol{x}$ is the input matrix for the images.

A pre-trained CNNs model is deployed for predicting the RUL of a cutting tool. The weights of the CNNs model are then adjusted through adaptively learning the images of cutting tools based on a transfer learning process. Meanwhile, to facilitate the CNNs model for the prediction, the end layers, which consist of the loss classifier and the classification output layer, are replaced with a designed regression layer stack as a regression problem.

More details of the developed approach are in the following steps: (1) forward computation of the CNNs is carried out by using the datasets of both the source domain (datasets for pre-training the CNNs) and the target domain (the images for tool heath state) as input for tool RUL prediction; (2) back propagation computation in the CNNs is performed to minimize the feature distribution discrepancy for the two domains and the prediction errors of the tool RUL. Gradient descent optimization

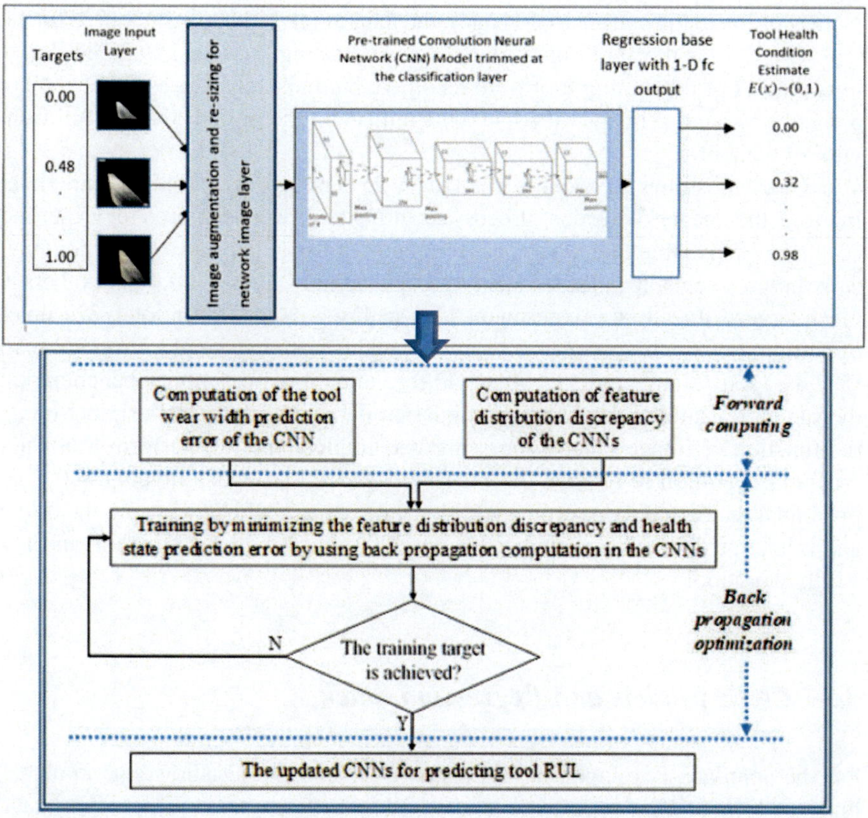

Fig. 1 The detailed procedure of the approach for tool RUL prediction

algorithms are evaluated and used for the minimization process, and the updated CNNs is deployed for tool RUL prediction. illustrates the above steps. More details on constituent components are given in the following sub-sections.

3.2 The Input Data for Transfer Learning

The dataset used for transfer learning comprises microscope images of the carbide cutting tool flank, collected at set intervals throughout the experiment, along with recorded flank tool wear width v in mm. The cutting experiments were conducted under varying conditions of cutting speed v_c, feed rate f_d, and cutting width a_e. In total, 25 experiments were used to vary these parameters following a Taguchi Design of Experiments (DoE) approach, with 3 factors (v_c, f_d, and a_e), 5 levels per factor, and 1 response variable (i.e., the measured flank wear width, v). The purpose is to analyze the effects of varying these parameters on the longevity of the tools.

In addition to the cutting tool images, the flank wear width was measured for each tool with a wear threshold of $v_b = 0.4$ mm indicating the tool has failed. Several image views of the cutting tool were recorded, but the analysis was focused on one particular variant of image views, at a magnification factor of 100 and a full frontal view of the tool.

Figure 2 illustrates the tool life trend of the 25 cutting tools within this experiment. In Fig. 3, the images of the post-processed cutting tools were captured for Experiment #1. In addition, some tool wear measurements (in particular those corresponding to early failure events, e.g., Experiment #15) had a final values $v < 0.4$ mm; others had much larger values (e.g., Experiment #21) with $v \sim = 1$ mm. In total, 327 images of cutting tools with appropriate tool wear width were used, split into 195 images (59.94%) for training and 132 (40.06%) for validation. To simplify benchmarking, the same pre-shuffled training and validation data were used. For original images, the function of image batch processing was implemented to perform a boundary cropping operation to remove the excess image backgrounds, yielding 800×800 pixel images. To avoid discarding additional data that could be relevant, the training and validation data were normalized between 0 (indicating a healthy tool) and 1 (for a fully worn tool).

3.3 CNNs Models and Regression Stack

For the approach developed in this paper, transfer learning allows the knowledge obtained from original tasks to be repurposed for different tasks. This means that the weights and biases of pre-trained CNNs models could be adjusted or fine-tuned with new training data. While the earlier feature pool layers of CNNs typically extract general image features that are considered safely transferable, specialized features are typically extracted in deeper layers. The degree of specialization often leads to some levels of pre-training bias, where the models retain features learned from the pre-training phase even despite being trained for extensive durations. It is intuitive to select models with a good performance in general classification to be further fine-tuned via transfer learning. This is because such a model would have been trained to accurately recognize features belonging to a multitude of different classes. Feature transferability is addressed using minimization optimization procedures for MMD and RUL prediction errors, described in Sect. 4 later on.

When selecting pre-trained CNNs models for transfer learning, another important consideration is the computational complexity of the models. While it is often observed that deeper models tend to outperform shallower ones at certain tasks, it is not always the case. The SqueezeNet architecture, for instance, was able to attain AlexNet-level accuracy on the ImageNet data challenge with nearly 50 times fewer parameters [34]. Therefore, comparing a variety of CNNs models quantitatively is useful to help evaluate their merits and appropriateness of this research. The CNNs models were chosen based on their classification performances in general-purpose

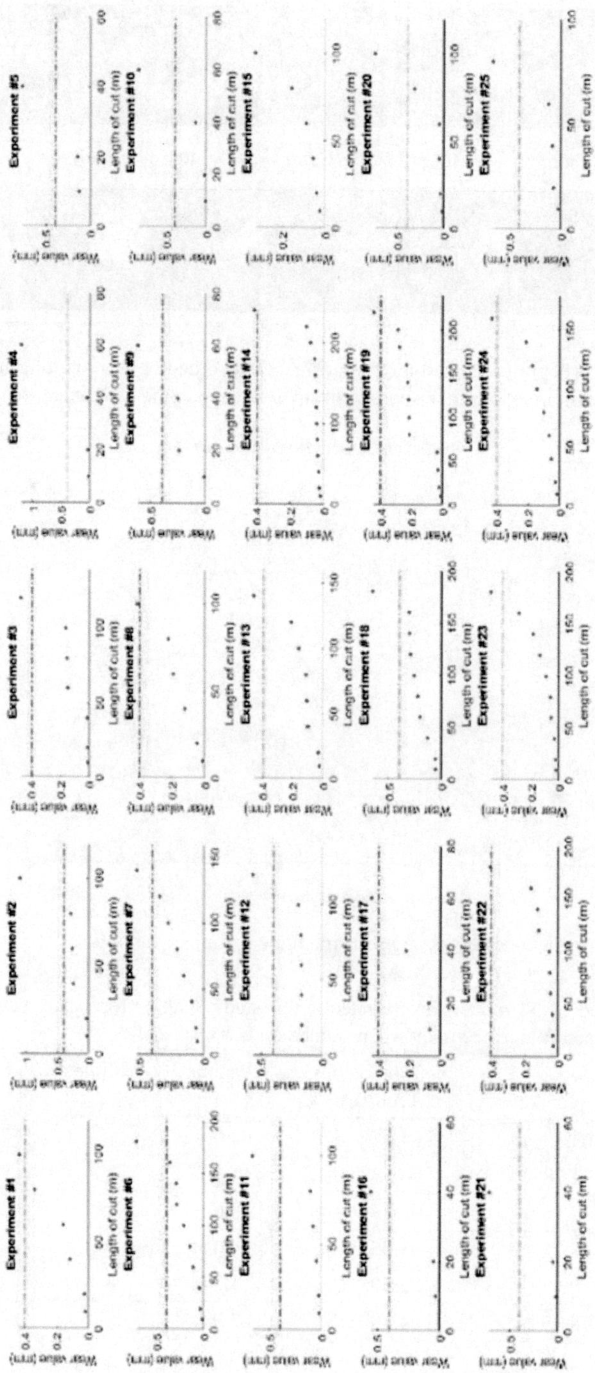

Fig. 2 Tool life trends of 25 cutting tool experiments

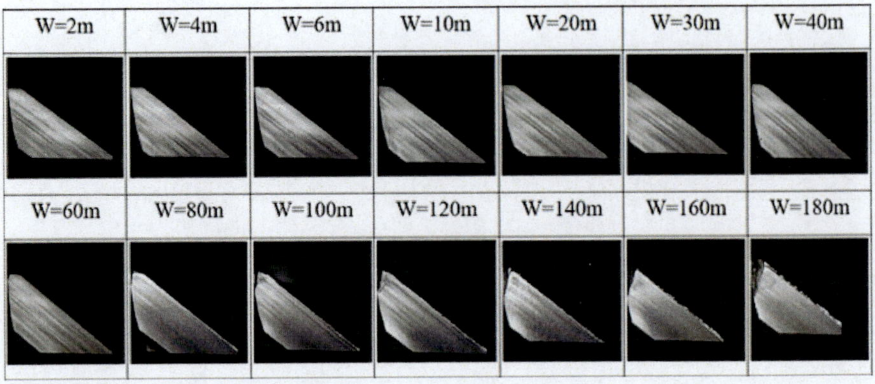

Fig. 3 Experiment #1: pre-processed tool wear images recorded at W m cutting intervals. The cutting interval in earlier measurements was kept small to capture the early wear trend

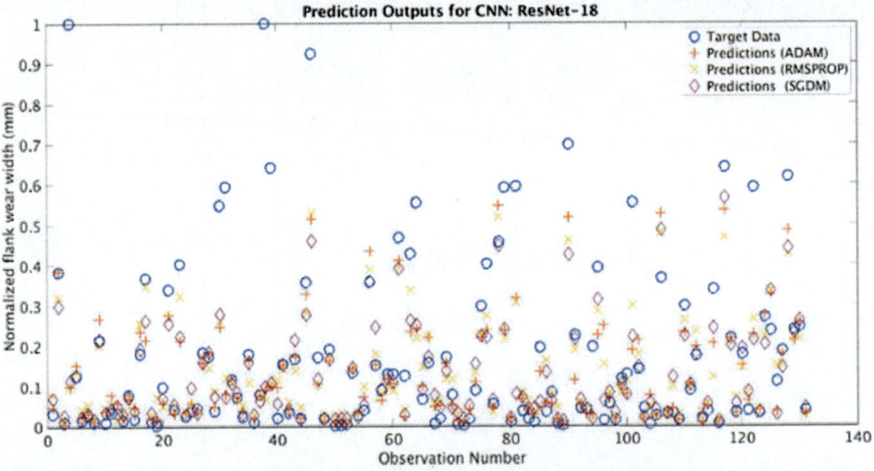

Fig. 4 ResNet-18 predictions versus targets using three optimizer variants

Table 1 Pre-trained CNNs models investigated in this study, with performance reported in terms of top-1 and top-5 percentage accuracy when trained on ILSVRC 2012

Network	Top-1 Accuracy (%)	Top-5 Accuracy (%)	Parameters (Millions)	Input Size
AlexNet [21]	63.3	84.6	61.0	227 × 227
ResNet-18	71.78	90.58	11.7	224 × 224
ResNet-50	77.15	93.29	25.6	224 × 224
ResNet-101	78.25	93.95	44.6	224 × 224
SqueezeNet	60.4	82.50	1.24	227 × 227
InceptionV3	78.95	94.49	23.9	299 × 299

classification tasks. Table 1 highlights the model size, input image size, and results from classification challenges for the CNNs models.

The regression stack is a collection of layers that progressively adapt the outputs of the pre-trained CNNs to make them more suitable for regression-based prediction. It comprises the following layers:

- A 4096-channel fully-connected layer, which function is to further down-sample the outputs of the previous pooling layer (which is a common design choice for CNNs models);
- A batch normalization layer, responsible for normalizing the inputs of the previous layer;
- Rectified Linear Unit (ReLU), which applies a non-linear transformation to the prior layer outputs;
- A 410 fully-connected layer, which down-samples the previous layer inputs;
- A sigmoid layer, which transforms the outputs of the previous layer to the range (0, 1) via the sigmoidal activation function;
- A regression output layer, which computes the loss of the prediction $\widehat{y}_i$.

As opposed to a classification layer that computes the probability of an image belonging to a given image class, the regression output layer computes the loss as the MSE (Mean Square Error) of the prediction $\widehat{y}_i$ given the target y_i. MSE is defined as:

$$MSE = \frac{1}{N}\sum_{i=1}^{N}\sqrt{(y_i - \widehat{y}_i)^2} \tag{2}$$

where N is the node number of the layer.

3.4 Transfer Learning Process

The CNNs model with the regression stack is re-trained on the training dataset of 195 images through the procedure illustrated in Fig. 1. In order to transfer the knowledge from the source domain (pre-trained CNNs) to the target domain (trained CNNs for tool RUL prediction), the developed approach should be subject to the condition that features are in similar distributions between domains. To address feature distribution mismatch during transfer learning, an error minimization optimization strategy is applied through back propagation computing on the pre-trained CNNs. In the previous literature, MMD (Maximum Mean Discrepancy) was popularly used to measure the distance metric for probability distribution between two domains. That is, the datasets in the source domain and the target domain are represented as $D_S = \{X_S, P(x_s)\}$ and $D_T = \{X_T, P(x_T)\}$ respectively. Meanwhile, $X_S = \prod_{i=1}^{ns}\{x_s^i, y_s^i\}$ with ns samples, and $X_T = \prod_{i=1}^{n_t}\{x_T^i\}$ with n_t samples respectively. Their MMDs are defined below:

$$Mean_H(X_S) = \frac{1}{n_s}\sum_{i=1}^{n_s} H(x_s^i) \quad (3)$$

$$Mean_H(X_T) = \frac{1}{n_t}\sum_{j=1}^{n_t} H\left(x_T^j\right) \quad (4)$$

$$MMD_H(X_S, X_T) = \sup[Mean_H(X_S) - Mean_H(X_T)] \quad (5)$$

where sup(·) represents the supremum of the aggregate; $H(\cdot)$is a RKHS (Reproducing Kernel Hilbert Space).

In this research, the above concept of MMD is adopted for measuring the feature distribution difference of domain invariant features. To achieve similar distributions from two domains, $MMD_H(X_S, X_T)$ is considered as the optimization objective to regularize the weights of the CNNs. Meanwhile, during the re-weighting process on the CNNs, the prediction error should be minimized as well. Thus, the prediction error is considered as another optimization objective.

As discussed earlier, the total loss function *Loss* can be calculated based on $MMD_H(X_S, X_T)$and MSE. Since $MMD_H(X_S, X_T)$ and MSE are in different value ranges, normalization is required. In this research, Nadir and Utopia points are utilized to normalize the above three objectives in an effective means [35]. The Utopia point ${z_i}^U$ provides the lower bound of No. *i* objective obtained by minimizing the objective as below:

$$ {z_i}^U = minf(i) \quad (6)$$

The Nadir point ${z_i}^N$ provides the upper bound of No. *i* objective by maximizing the objectives:

$$ {z_i}^N = \max_{1 \ll j < I} f(j) \quad (7)$$

where *I* is the total number of the objective functions.

According to Eqs. (5) and (6), the normalized $MMD_H(X_S, X_T)$and MSE can be calculated below:

$$NMMD_H = (MMD_{H1}(X_S, X_T) - {z_1}^u)/({z_1}^N - {z_1}^u) \quad (8)$$

$$Nmse = (MSE - {z_2}^u)/({z_2}^N - {z_2}^u) \quad (9)$$

where $NMMD_H$ and $Nmse$ are the normalized $MMD_H(X_S, X_T)$and MSE respectively.

The total loss function *Loss* can be calculated based on the weighted sum of the two normalized objectives:

Table 2 Fine-tuning the transfer learning enabled CNNs using different optimizers

1. Imagesize = size of image input layer
2. Identify the last (tune-able) layers in the network
3. Set learning rate to 0 for other layers
4. For each optimizer
5. If optimizer is ADAM or RMSProp
6. Set initial learning rate to 4e-5
7. Else if optimizer is SGDM
8. Set initial learning rate to 2e-2
9. Pre-initialize augmenter which augments the input images by resizing and performing random transformations
10. Train the network on augmented images
11. End

$$Loss = w_1 \cdot NMMD_H + w_2 \cdot Nmse \quad (10)$$

where $w_1 - w_2$ are the weights of the two objectives, and $\sum_{i=1}^{2} w_i = 1$.

Based on the above process, three variants of training optimization algorithm were investigated and compared, including Stochastic Gradient Descent with Momentum (SGDM), Root Mean Square Propagation (RMSPROP) and Adaptive Moments (ADAM) [36]. SGDM has been a popular choice for training ANNs since its inception in 1999, and its subsequent resurgence when used in AlexNet. RMSProp is another popular algorithm for gradient descent training to eliminate the need for learning rate adjustment. ADAM combined the heuristics of both Momentum and RMSProp to achieve faster convergence.

The CNNs models were trained according to the procedure illustrated in Table 2, after being converted to a *layerGraph* (the MATLAB structure that retains the CNNs configuration). Firstly, a helper function searches the *layerGraph* of the CNNs model for a Softmax layer, its preceding layer (a fully-connected layer with 1000 outputs) and subsequent classification layer. The function returns the names and indices of these layers in the *layerGraph*, after which they are removed using a helper function, and replaced with the *baseLayers* configuration described in the regression stack.

The models were trained for 750 epochs in 12 iterations per epoch (9000 iterations total), with a mini-batch size of 16 images. The training function was set to shuffle the mini-batch every epoch. To speed up training, validation was done every 40 iterations. During training, image augmentation operations were implemented on each training mini-batch, to vary the input images by introducing some aspects of visual variation to the images (random translations, rotations, and scaling). This has the effect of inflating the dataset, allowing the CNNs model to consider more examples of the data than are available.

4 Experimental Results and Discussions

4.1 Experiment Data Collection

Table 3 details the benchmarking results for fine-tuning the 6 CNNs models by the transfer learning process, where the 3-run average of each model variant's output was recorded. To evaluate the quality of prediction, the models were assessed with the following performance criteria:

(1) Training time (in seconds).
(2) Mean Absolute Error (MAE), which is defined below:

$$MAE = \frac{1}{N}\sum_{i=1}^{N}(y_i - \hat{y}_i)^2 \tag{11}$$

(3) Mean Square Error (MSE), which has been defined in Formula (2).
(4) $acc_{10,20,30}$, the accuracies of all predictions are below 10%, 20% or 30% error thresholds from the targets. The threshold of the T percentage accuracy is:

$$acc_T = \frac{1}{N}\sum_{i=1}^{N} 1_T((\hat{y}_i)) \tag{12}$$

$$1_T(\bar{y}_i) := \begin{cases} 1, (\hat{y}_i \geq T \times |\max(y_i)| \\ 0otherwise \end{cases} \tag{13}$$

where the range of T is determined by $T \in [0.1, 0.2, 0.3]$.

The threshold of 10% accuracy indicates the percentage of prediction that falls within 10% of this error in either direction. In most cases, due to the proportion of healthy samples compared to faulty or worn tool images, the prediction errors are on the lower end of the range, i.e., ~10% below the target value. The performance measures of 20% and 30% accuracy are considered additional qualitative metrics indicating whether predictions can be accepted.

4.2 Result Analysis and Observations

In Table 3, comparing these results, ResNet-18 (trained with ADAM) can offer the best performance in terms of average prediction error and acc10, with a reasonable training time considering the number of iterations attempted. ResNet-50 and ResNet-101 variants are both longer to train and generally less accurate based on MAE and MSE. Meanwhile, despite being much deeper than other models, the InceptionV3

Table 3 Benchmarking results for fine-tuning the 6 CNNs models

Pre-trained Model	Training Details			Model Performance					
	Optimizer	Learning Rate	Batch Size	MAE	MSE	acc10 (%)	acc20 (%)	acc30 (%)	Training Time (s)
AlexNet [34]	adam	4e-5	16	0.0829	0.1684	81.68	**90.84**	91.60	2124.8
	sgdm	2e-2	16	0.0868	0.1723	79.39	90.08	91.60	1940.1
	rmsprop	4e-5	16	0,0903	0.1726	77.86	90.08	90.84	2027.6
ResNet-18 [31]	adam	4e-5	16	**0.0773**	0.1654	**83.97**	**90.84**	92.37	3358.4
	sgdm	2e-2	16	0.0820	0.1591	78.63	90.08	92.37	2649.0
	rmsprop	4e-5	16	0.0791	0.1594	80.15	90.08	92.37	2853.5
ResNet-50 [31]	adam	4e-5	16	0.0868	0.1764	80.92	86.26	90.84	14,790
	sgdm	2e-2	16	0.1124	0.1967	74.05	84.73	90.08	9184.2
	rmsprop	4e-5	16	0.1050	0.1954	76.34	83.97	90.08	12,689
ResNet-101 [31]	adam	4e-5	16	0.0833	0.1657	74.81	89.31	92.37	17,750
	sgdm	2e-2	16	0.0992	**0.1565**	74.05	89.31	**93.89**	15,122
	rmsprop	4e-5	16	0.0882	0.1740	77.86	86.26	90.84	16,654
SqueezeNet [37]	adam	4e-5	16	0.0891	0.1732	79.39	88.55	91.60	2151.8
	sgdm	2e-2	16	0.0868	0.1784	79.39	89.31	91.60	**1763.5**
	rmsprop	4e-5	16	0.0882	0.1710	77.68	88.55	91.60	1878.0
InceptionV3 [32]	adam	4e-5	16	0.0886	0.1784	79.39	85.50	91.60	20,334
	sgdm	2e-2	16	0.1040	0.1931	77.10	85.50	90.08	14,653
	rmsprop	4e-5	16	0.0916	0.1728	79.39	87.02	91.60	18,780

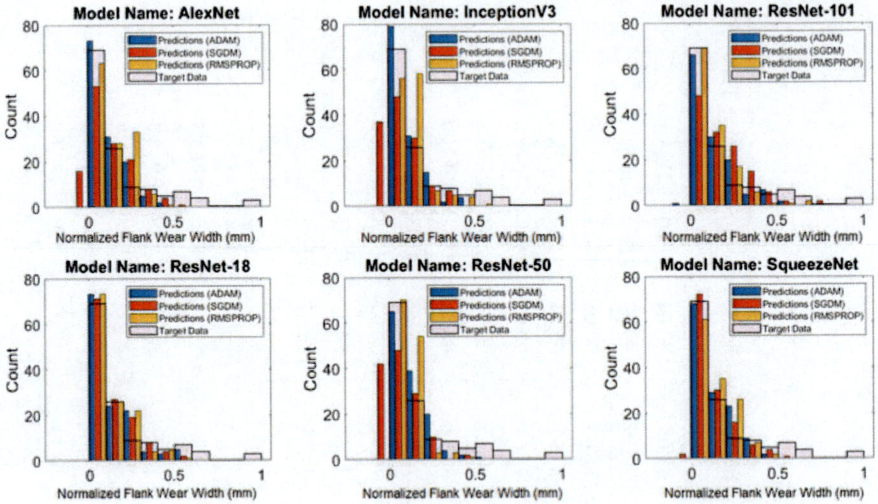

Fig. 5 Prediction histogram comparison between six benchmarked CNNs models

variants were amongst the worst performing models considering MAE, MSE and the accuracy thresholds. Furthermore, the increase in model depth from ResNet-18 to ResNet-101 has increased training time fivefold, without an improvement in performance. This emphasizes a key observation that increase of the model depth does not mean increase in prediction accuracy accordingly. Sample prediction outputs using the three optimizers chosen (ADAM, RMSPROP and SGDM) are illustrated in Fig. 4.

Figure 5 shows a histogram plot of the prediction outputs of the six CNNs. Some further observations can be made regarding the performance of these models:

- Overall best fit: It shows that ResNet-18 produced the closest prediction output distribution to the validation target data, across the three optimizer training variants.
- Overfitting: All models over-fit the results significantly in the "healthy" categories, with the performance of ResNet-18 (ADAM) being the best out of the compared model variants in terms of overfitting, where the less the model over-fits, the better its performance.
- Generalization performance: Comparatively, ResNet-50 produced the worst general fit results, indicated by its comparatively higher MAE and MSE as well as lower accuracy across all thresholds. This might indicate that the model has a tendency to over-fit the data more strongly than other models. In fact, the generalization performance of SqueezeNet, which is close to 20 times smaller in parameters, is markedly better consider the relative difference in model size.
- Anomalous predictions: With the exception of ResNet-18, all models trained with SGDM have a tendency to produce negative outputs, despite the sigmoid layer (whose function is to force its outputs to be between 0 and 1) being the last layer

prior to the regression output layer. This is a property of SGDM which enables it to generalize better than the other training algorithms. However, in doing so the SGDM variants predict results in the reverse direction of what is desired. This contrasts to tool wear width values, which must always be indicated by a positive value.

- Training duration: It is also worth mentioning that increasing the number of epochs to 9000 did not have a profound impact on the accuracy. Some initial trials with fewer iterations (i.e., 150 instead of 750) yielded similar results for most of the models. It is common to select a short training duration for the fine-tuning process.

Figures 6, 7, 8, 9 compare the results (accuracy, log(training time), MAE, and MSE) from all the model variants (ADAM, SGDM and RMSPROP). ResNet-18 is clearly shown to have the highest average accuracy and lowest MAE, despite being slightly longer to train than AlexNet in training time. ResNet-18 is also amongst the best performing models for MSE, bested only by ResNet-101 trained with SGDM. It therefore concludes that ResNet-18 is the best performed CNNs at learning a new task (regression output of normalized tool wear state) from images of tools using transfer learning.

From the above analysis, the prediction workflow of tool health state based on transfer learning enabled ResNet-18 variants can be more effective in early stages (i.e., good tool health). However, data imbalance and overfitting have considerable negative impacts on prediction accuracy, where classes are not uniformly distributed across the dataset. This is evident in the collected data in this research, where many more examples of healthy tools (i.e., $v < 0.4$) are available than those of less healthy tools ($v \geq 0.4$). This is made further apparent in Fig. 10 that shows the distribution of the normalized training and testing dataset targets; there are much fewer values

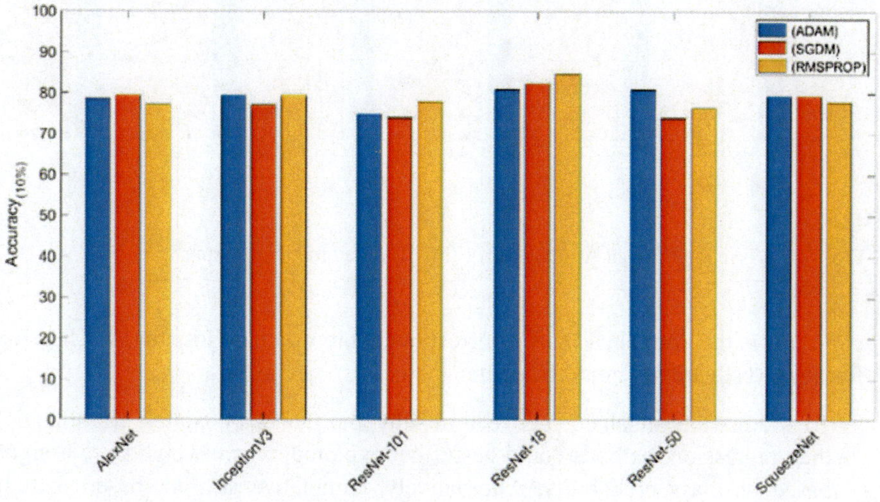

Fig. 6 CNNs models' accuracy for all training variants

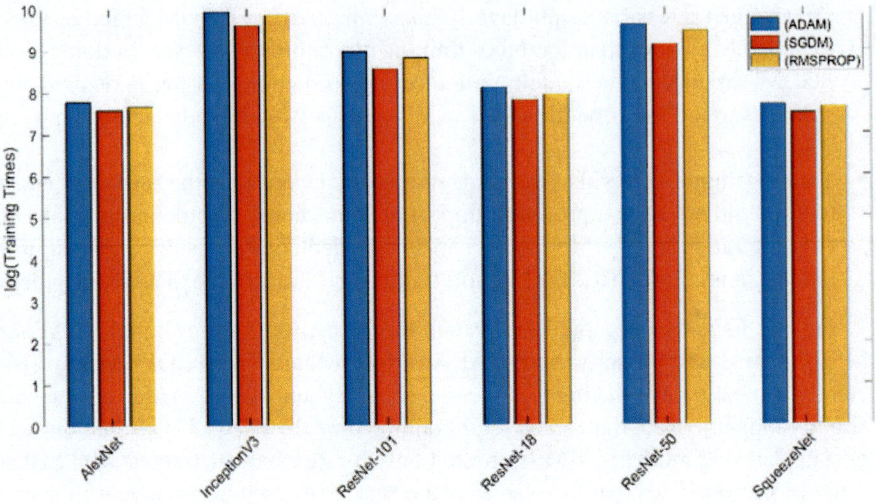

Fig. 7 CNNs models' log (training times) for all training variants

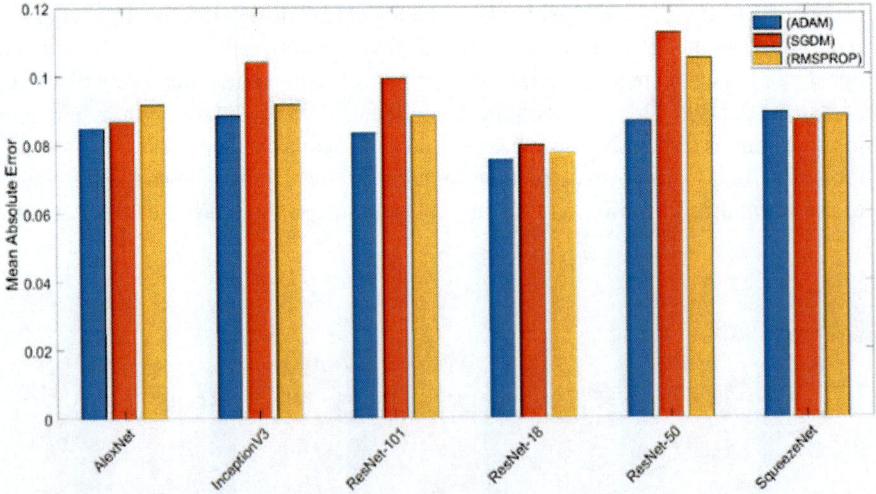

Fig. 8 CNNs models' Mean Absolute Error (MAE) for all training variants

close to 1 in the normalized scale, corresponding to v values close to 0.4. Therefore, further investigations should be made:

- To address the imbalance between healthy and faulty tool states, classification-then-regression methods could be further explored, where weights are assigned based on class probability. Alternatively, cumulative attribute-based methods could help improve accuracy by reformulating the regression problem in a manner similar to classification. Another alternative could be explored based on parameter

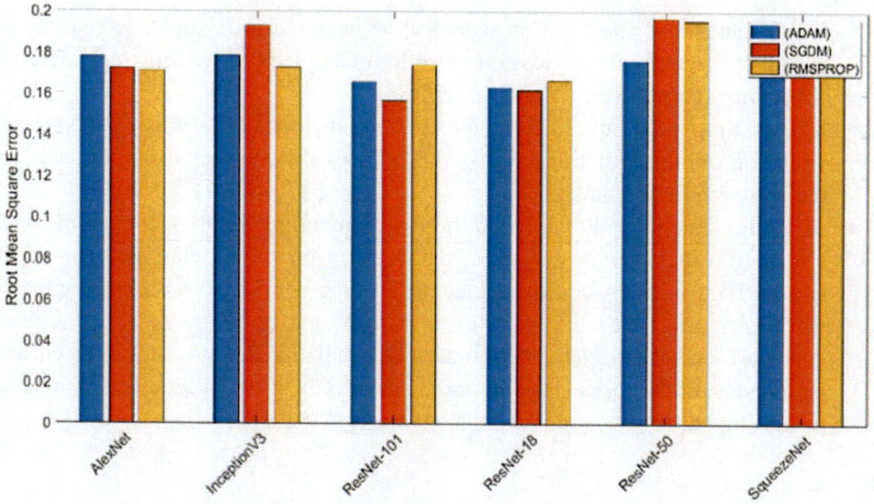

Fig. 9 CNNs models' Mean Square Error (MSE) for all training variants

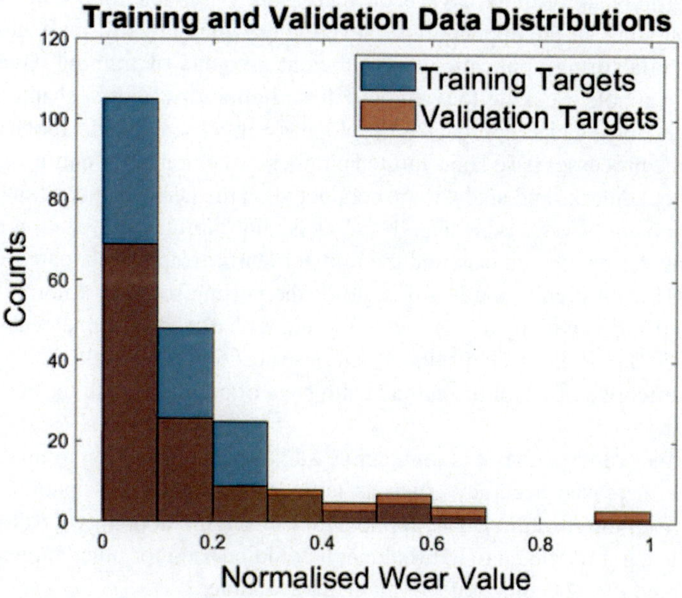

Fig. 10 The data distribution of the training and validation target data

transfer approaches, where a source task (and its corresponding source domain data) is used to pre-train the model.

- Additional works are required to improve the accuracy of prediction across increasing wear levels (i.e., where the normalized wear value exceeds 0.5). Some additional pre-manipulations of the data need to be implemented, by adding extra safety margins to the hand-measured wear values, for example. Increasing the cost parameters for the regression layer, for example by increasing regularization L2-norm penalties, could reduce overfitting.

- Investigating maximum likelihood estimation methods for regression could help with improving predictions across the full range of expected outputs, thereby reducing prediction bias.
- As a supplement or alternative to these aforementioned approaches, some techniques to re-balance the classes of datasets, or implement class penalties using target RUL functions, could enhance the accuracy of the CNNs-based regression workflow.
- An end-to-end approach to incorporate additional model inputs such as machining parameters or categorical labels, combined with CNNs tool wear estimation, could be developed.

5 Conclusions

Deep learning algorithms have been increasingly applied for PHM due to their great potentials in the applications. Nevertheless, they are still ineffective in practical manufacturing applications as sufficient amounts of training dataset are not usually available. Seeking to overcome these limitations, in this chapter, a transfer learning enabled CNNs approach is developed to effectively predict tool RUL in CNC machining processes based on a limited number of the images of cutting tools. Quantitative benchmarks and analysis are conducted on the performance of the developed approach using several typical CNNs models and training optimization techniques. Experimental results indicate that the transfer learning approach, particularly using ResNet-18, can predict the health state of the cutting tool (as a normalized value between 0 and 1) with up to 84% accuracy and with a prediction mean absolute error of 0.0773. Based on these results, it demonstrates that the developed approach can achieve effective predictions on the health state of cutting tool in the early stages of tool wear.

A further research work is to integrate additional information to predict the tool RUL for increased accuracy (such as temperature, power dissipation, or current signals from the machine). The applicability of the methodology developed in this approach is not restricted to PHM alone; it could be used for other applications with only limited datasets in a target domain are available.

References

1. Compare M, Bellani L, Zio E (2017) Reliability model of a component equipped with PHM XE "PHM" capabilities. Reliab Eng Syst. Saf 168:4–11
2. Lu B, Zhou X (2017) Opportunistic preventive maintenance scheduling for serial-parallel multistage manufacturing systems with multiple streams of deterioration. Reliab Eng Syst Saf 168:116–127
3. Li X, Zhang W, Ding Q (2019) Deep learning XE "Deep learning" -based remaining useful life estimation of bearings using multi-scale feature extraction. Reliab Eng Syst Sa. 182:208–218
4. Oh JW, Jeong J (2019) Convolutional neural network and 2-D image based fault diagnosis of bearing without retraining. Proceedings of the 3rd international conference on compute and data analysis - ICCDA 2019, pp 134–138
5. Hoang DT, Kang HJ (2019) Rolling element bearing fault diagnosis using convolutional neural network and vibration image. Cogn Syst. Res 53:42–50
6. Zhao R, Yan R, Wang J, Mao K (2017) Learning to monitor machine health with convolutional Bi-directional LSTM XE "LSTM" networks. Sensors 17(2):273
7. Ghani JA, Rizal M, Nuawi MZ, Ghazali MJ, Haron CHC (2011) Monitoring online cutting tool wear using low-cost technique and user-friendly GUI. Wear 271(9–10):2619–2624
8. Lei Y, Li N, Guo L, Li N, Yan T, Lin J (2018) Machinery health prognostics: A systematic review from data acquisition to RUL XE "RUL" prediction. Mech Syst Signal Process 104:799–834
9. Johansson D, Hägglund S, Bushlya V, Ståhl JE (2017) Assessment of commonly used tool life models in metal cutting. Procedia Manuf. 11:602–609
10. Zhao R, Yan R, Chen Z, Mao K, Wang P, Gao RX (2019) Deep learning XE "Deep learning" and its applications to machine health monitoring. Mech Syst Signal Process 115:213–237
11. Deng J, Dong W, Socher R, Li L-J, Li K, Li FF (2010) ImageNet: A large-scale hierarchical image database. Proceedings of the 2009 IEEE conference on computer vision and pattern recognition, pp. 248–255.
12. Russakovsky O et al (2015) ImageNet large scale visual recognition challenge. Int J Comput Vision 115:211–252
13. Krizhevsky A, Nair V, Hinton GE (2020) CIFAR-10 and CIFAR-100 datasets. https://www.cs.toronto.edu/~kriz/cifar.html. Accessed 10 Jan 2020
14. Janssens O, Van De Walle R, Loccufier M, Van Hoecke S (2018) Deep learning XE "Deep learning" for infrared thermal image based machine health monitoring. IEEE/ASME Trans Mechatronics 23(1):151–159
15. Pan SJ, Yang Q (2010) A survey on transfer learning. IEEE Trans Knowl Data Eng 22:1345–1359
16. Weiss K., Khoshgoftaar T.M., Wang D.D., 2016. A survey of transfer learning. J. Big Data. 3(1).
17. Gouarir A, Martínez-Arellano G, Terrazas G, Benardos P, Ratchev S (2018) In-process tool wear prediction system based on machine learning techniques and force analysis. Procedia CIRP 77:501–504
18. Hsu C-SS, Jiang J-RR (2018) Remaining useful life estimation using long short-term memory deep learning. Proceedings of the 2018 IEEE international conference on applied system invention (ICASI), pp 58–61
19. PHM Society (2010) 2010 PHM society conference data challenge. https://www.phmsociety.org/competition/phm/10. Accessed 15 Jan 2020
20. Günther J, Pilarski PM, Helfrich G, Shen H, Diepold K (2014) First steps towards an intelligent laser welding architecture using deep neural networks and reinforcement learning. Procedia Technol 15:474–483
21. Saxena A, Goebel K, Simon D, Eklund N (2008) Damage propagation modelling for aircraft engine run-to-failure simulation. Proceedings of the 2008 international conference on prognostics and health management (PHM 2008), pp. 1–9.
22. Lu C, Wang Z, Zhou B (2017) Intelligent fault diagnosis of rolling bearing using hierarchical convolutional network based health state classification. Adv Eng Inform 32(139–151)

23. Li Z, Wang Y, Wang K (2019) A deep learning driven method for fault classification and degradation assessment in mechanical equipment. Comput Ind 104:1–10
24. Zhang J, Wang P, Yan R, Gao RX (2018) Learning improved system remaining life prediction. Procedia CIRP 72:1033–1038
25. Shi C, Panoutsos G, Luo B, Liu H, Li B, Lin X (2018) Using multiple feature spaces-based deep learning for tool condition monitoring in ultra-precision manufacturing. IEEE Trans Ind Electron 66:1–1
26. Liu X, Li Y, Chen G (2019. Multimode tool tip dynamics prediction based on transfer learning. Robotics and Computer Integrated Manufacturing. 57: 146–154.
27. Zhou B, Khosla A, Lapedriza A, Torralba A, Oliva A (2016) Places: an image database for deep scene understanding. https://arxiv.org/abs/1610.02055. Accessed 08 Jan 2019
28. Lu W, Liang B, Cheng Y, Meng D, Yang J, Zhang T (2017) Deep model based domain adaptation for fault diagnosis. IEEE Trans Ind Electron 64(3):2296–2305
29. Xiao D, Huang Y, Zhao L, Qin C, Shi H, Liu C (2019) Domain adaptive motor fault diagnosis using deep transfer learning. IEEE Access 7:80937–80949
30. Wen L, Gao L, Li X (2019) A new deep transfer learning based on sparse auto-encoder for fault diagnosis. IEEE Trans Syst, Man, Cyber: Syst 49:136–144
31. He K, Zhang X, Ren S, Sun J (2015) Deep residual learning for image recognition. Proc IEEE Conf Comput Vis Pattern Recognit 770–778.
32. Szegedy C et al (2015) Going deeper with convolutions. Proceedings of the IEEE computer society conference on computer vision and pattern recognition, pp 1–9
33. Qian N (1999) On the momentum term in gradient descent learning algorithms. Neural Networks 12(1):145–151
34. Tieleman T, Hinton GE, Srivastava N, Swersky K (2012) Lecture 6.5-rmsprop: Divide the gradient by a running average of its recent magnitude. COURSERA Neural Networks Mach. https://www.cs.toronto.edu/~tijmen/csc321/slides/lecture_slides_lec6.pdf. Accessed 10 Jan 2019
35. Mausser H (2019) Normalization and other topics in multi-objective optimization. Proceedings of the fields–MITACS industrial problems workshop, pp 59–101
36. Ruder S (2016) An overview of gradient descent optimization algorithms. https://arxiv.org/abs/1609.04747. Accessed 10 Jan 2019
37. Iandola FN, Han S, Moskewicz MW, Ashraf K, Dally WJ, Keutzer K (2016) SqueezeNet: AlexNet-level accuracy with 50x fewer parameters and <0.5MB model size. Proc Int Conf Comput Vis Pattern Recognit

Prediction of Strength of Adhesive Bonded Joints Based on Machine Learning Algorithm and Finite Element Analysis

Y. C. Liang, Y. D. Liu, and W. D. Li

1 Introduction

Aerospace, vehicle and civil engineering sectors have witnessed a quick swift from using single materials to multi-materials to develop components and structures to fulfil sophisticated functions [1]. To bond multiple thin layers of different materials into a component (a substitute), there is a need to develop effective adhesive joining technologies that can exhibit excellent stress transfer behaviours and compatibility with a wide range of material types [2]. For an adhesive joining technology, the failure load is critical to determine joint performance and important parameters, such as elastic modulus and fracture parameters of substitutes. Therefore, it is critical to effectively predict the maximum failure load and identify optimal parameters. Joints' failure loads are usually acquired through lap shear tests on bundles of joint coupons. However, though the results of lap shear tests are usually reliable, experimental tests are highly sensitive to the material category of adherends and adhesive, making the experiments costly or even impractical to be conducted for each category of adherend and adhesive. To predict the failure load of joints and to reduce the number of experiments, the Finite Element Analysis (FEA) method has become a popular method. The advantages of FEA are: (i) boundary value problems can be resolved systematically; (ii) components with complex geometry can be resolved more accurately using FEA in comparison with experiments [3]; (3) a FEA model just needs a single lap testing experiment for validation so that the number of required experiments is

Y. C. Liang · W. D. Li (✉)
Faculty of Engineering, Environment and Computing, Coventry University, Coventry, UK
e-mail: weidong.li@coventry.ac.uk

Y. D. Liu
Warwick Manufacturing Group (WMG), University of Warwick, Coventry, UK

W. D. Li
School of Logistics Engineering, Wuhan University of Technology, Wuhan, China

W. Li et al. (eds.), *Data Driven Smart Manufacturing Technologies and Applications*, Springer Series in Advanced Manufacturing,
https://doi.org/10.1007/978-3-030-66849-5_8

reduced dramatically. However, the challenging issue in the FEA method is that FEA simulation is computational expensive, and the correlation between parameters in the FEA model is not easy to be disclosed, making the selection of optimal parameters difficult.

In order to tackle the challenge, in this chapter, an innovative approach based on machine learning and FEA is developed to predict the maximum failure load. The innovations are stated as follows:

- The prediction of strength of bonded component joints is achieved based on a Deep Neural Networks (DNNs), which is trained by using datasets generated through the FEA model. Case studies demonstrated that the approach is more effective than the prediction approach just using a FEA model, and 99.54% less computational time was consumed;
- The machine learning algorithm, i.e., DNNs is designed to represent customized failure load modelling flexibly. The structure and parameters of DNNs are justified in order to ensure the best accuracy and efficiency.
- Based on the computation of predictive approach, an innovative optimization algorithm, i.e., Fruit Fly Optimization algorithm (FFO), is developed to efficiently identify the optimal material parameters with a given geometry.

2 Literature Survey

To build a model to predict the static strength of an adhesive joint, numerous analytic modelling methods were developed, including the Ojalvo–Eidinoff approach [4], the Volkersen's approach [5], the Goland-Reissner approach [6], the Shear-Lag approach [7], etc. The analytic approaches, however, are not accurate in complicated cases. Thus, these analytical approaches have been recently replaced by a numerical analysis approach, such as Finite Element Analysis (FEA). In these aspects, there are a number of investigations conducted. Bahrami et al. [8] designed a FEA model to predict the failure load of adhesive joints based on different parameters, such as the notch angle, the notch width, the notch depth, and the notch distance from the overlap length. Sadeghi et al. [9] utilized some FEA models to predict the failure behaviors of single lap joints (SLJ) based on fracture toughness criterion, and the failure load was predicted with high accuracy compared with experiment results. Saleh et al. [9] introduced FEA to predict the adhesive shear stress/strain distribution of double-lap Steel-to-CFRP adhesively bonded joints loaded in tension. Ramalho et al. [10] designed a meshless FEA model to predict the strength of SLJ for three adhesives. The aforementioned research works have shown a good accuracy of prediction for adhesive strength.

However, FEA simulation is highly computational expensive. In order to reduce the computational cost of a FEA model or expensive laboratory experiments, machine learning based approaches have been increasingly explored to reduce the need of FEA in various engineering problems. Chou et al. [11] compared four different intelligent algorithms, including multilayer perceptron (MLP) neural network, support vector machine (SVM), classification and regression tree (CART), and linear regression (LR), for concrete strength prediction. The algorithms were tested with five reliable datasets from experimental tests, and SVM and MLP showed good prediction performance. Zhu and Zhang [12] proposed a machine learning algorithm with long-term field measurements and large-scale data from multi-scale FEA. In the research, a case study for a prototype cable-stayed bridge was used and the computational cost was significantly reduced. Saadallah et al. [13] designed a machine learning algorithm to achieve a monitoring purpose during machining processes with reliable prediction of process reliability. The trained data was generated through physical simulation. Capuano and Rimoli [3] developed a regression model-based approach to calculate the relationship between the element state and its forces. The advantage of using the approach is that the computational cost can be significantly reduced by avoiding finding the internal displacement field and eliminating the requirement of numerical iterations. Al-Shamiri et al. [14] designed extreme learning machine (ELM) to predict the compressive strength of high-strength concrete (HSC) based on 324 experiment data records from laboratory. The accuracy achieved over 98% using the ELM approach. For the above research works, most of researchers used experimental data to train machine learning-based approaches, while experiments for adhesive strength are expensive and time consuming for setup and conducting. Therefore, it is imperative to develop a systematic approach to use FEA simulation data that are justified by experiment data to train a machine learning algorithm to build a predictive model. However, to the best of our knowledge, there is rare research in this aspect of prediction of strength of adhesive bonded joints.

Once a predictive model is built, it is beneficial to design an optimization algorithm to find the mostly proper parameters to ensure the best material property. Some research works have been developed to design and implement optimization algorithms for the purpose of material property improvement. Putra et al. [15] proposed a hybrid Genetic Algorithm (GA) technique to reduce weight of ship to ensure the minimum CO_2 emission. The designed variables are type of stiffeners, stiffener spacing, and plate thickness. The designed hybrid GA is able to handle the discrete variable such as type of stiffeners. Wu et al. [16] proposed a globally convergent version of the method of moving asymptotes (GCMMA) to optimize the multi-material topology for thermomechanical buckling problems. Chegini et al. [17] proposed a particle swarm optimization algorithm (PSO) based on Sine Cosine algorithm and levy flight to improve the performance of original PSO, so that the problem of PSO for being trapped into local optima can be avoided. Tsiptsis et al. (2019) proposed Particle Swarm Optimizer (PSO) to solve size, shape and topology optimization problems of both 2D truss and frame towers. Javidrad et al. [18] proposed a hybrid particle swarm optimization (PSO) and simulated annealing (SA) methods to reduce coupling effects and improve stiffness and strength of the laminate. Layer thickness and fibre orientation angle of each ply group are defined as variable. However, there is still limitation for the current research: (i) few research has considered integrating optimization algorithms with machine learning algorithms for material strength optimization, (ii) it is crucial to select the suitable and computationally efficient optimization algorithm for optimal parameter selection.

3 System Structure

Figure 1 shows the overall system architecture of the machine learning and FEA enabled approach for prediction of strength of adhesive bonded joints and parameter optimization. The details of the system are explained below:

- Experiment stage: For the system, experiments were setup for lap shear testing of aluminium joints bonded with epoxy adhesive. The load–displacement behavior of joints was collected and used for FEA validation. The experiment data is based on a small sample size to justify the difference between experiment data and those generated from the FE model.
- FEA model stage: The validated FEA model can be used to simulate the failure load and maximum extension with different joint parameters. Based on this, a large amount of data samples can be collected for DNN training. The parameters used in the FEA modelling include E1 (elastic modulus of the upper adherends), E2 (elastic modulus of the lower adherends), G1 (mode I fracture toughness of adhesive) and G2 (mode II fracture toughness of adhesive).
- Machine learning modelling stage: The designed DNN model can be trained based on collected data, including both joint parameters as input and the failure load as output. The trained DNN model can be used to predict failure load based on given parameters. Based on the machine learning-based prediction model, a Fruit Fly optimization algorithm is developed to find the optimal parameter to achieve best failure load and maximum extension.

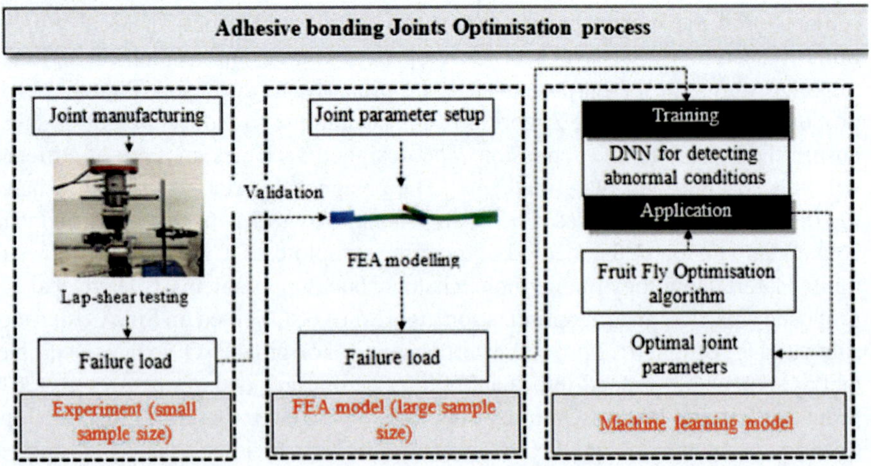

Fig. 1 Overall system structure

4 Detailed System and Algorithm

4.1 Experiment Stage

The geometry and dimensions of single lap joints are adopted from ASTM D3165 standard (2014), as shown in Fig. 2. The width of the joint was designed as 25 mm to tolerate loads. The adherends were manufactured from the aluminium alloy 6110-T6. The adhesive used is Araldite® 2015, which is a two-component, room temperature curing structural adhesive providing a resilient bond [19]. The nominal thickness of the adherend and adhesive is 2 mm and 0.25 mm, respectively. The surfaces of substrates were treated by abrasive sanding using P40 sandpaper followed by a cleaning process with ethanol solvent. A thin layer of Sika Primer 204 N was applied to favor the chemical bond at the adherend-adhesive interface [20]. The adhesive thickness was controlled by using glass beads of a given diameter. The joint specimens were cured at 120 °C for 40 min by using a clamping jig with fasteners to ensure a uniform adhesive layer. Aluminium end tabs were prepared to reduce the bending moment and to avoid damage to the adherends during the tests. The Young's modulus and Poisson's ratio of the aluminium is 70 GPa and 0.3, respectively. The material properties and cohesive parameters of this adhesive were measured by Campilho et al. [21], as shown in Table 1.

Lap shear testing: Lap shear tests of the single lap joints were carried out using an Instron 5500R machine with a 30 kN load cell, wedge-action grips and a displacement rate of 1 mm/min at room temperature. All the specimens were loaded under displacement control. The load and crosshead displacement were reset to zero-position after

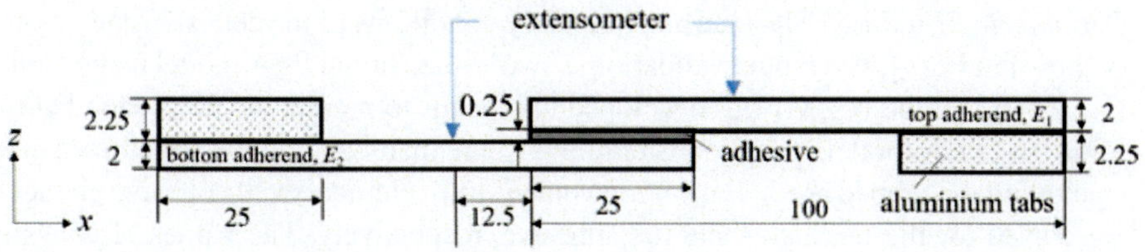

Fig. 2 Geometry and dimensions of single lap joints. The width of the joints is 50 mm. Unit in mm (ASTM D3165, 2014)

Table 1 Properties of the adhesive Araldite® (Huntsman, 2015)

Property	Araldite 2015
Young's modulus, E (GPa)	1.85 ± 0.21
Poisson ratio, ν	0.33
Tensile failure strength, σf (MPa)	21.63 ± 1.61
Shear failure strength, τf (MPa)	17.9 ± 1.8
Toughness in tension, G_n^c (N/mm)	0.43 ± 0.02
Toughness in shear, G_s^c (N/mm)	4.70 ± 0.34

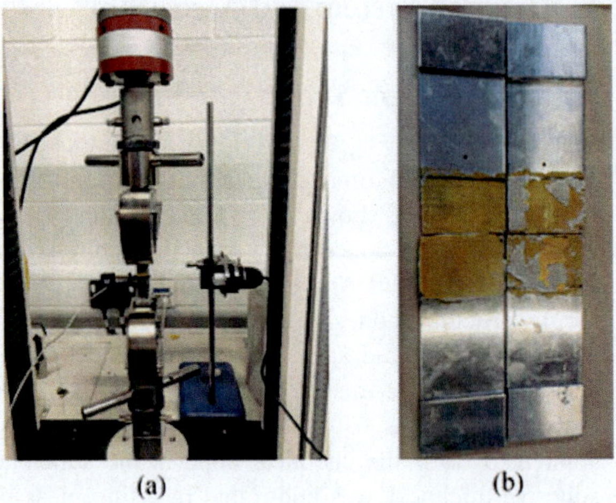

Fig. 3 **a** The locations of the reflective pads attached to the single lap joint specimens (unit in mm), **b** the setup of the test machine with laser extensometer

a small pre-loading. The mechanical extensometer was mounted at a defined distance of 50 mm to measure the relative displacement between the two pads attached to the bonded overlap area of the joints, as shown in Fig. 3a. The entire setting up of the test machine is shown in Fig. 3a. The joints were found in cohesion failure or mixed failure, as shown in Fig. 3b.

4.2 FEA Model Stage

Parametric FEAstudy: The load and boundary conditions of the cohesive zone model is shown in Fig. 4. For model validation, a two-dimensional FEA model using Cohesive Zone Method (CZM) was developed in Abaqus to predict the failure load of the joint. The FEA model of lap joints employs the realistic geometries and dimensions (including the tabbed area). Four-noded plane strain elements and cohesive elements were used for the laminates and the adhesive, respectively. The values of cohesive parameters are summarized in Table 2 [20]. The displacements and rotations of the entire nodes at one end were restrained in all directions, whereas those at the other end are loaded till final failure. A mesh size of 0.2 mm was used at the bonded area after a mesh convergence study and gradual mesh size is used for the laminates to save computational time. Non-linear geometric analysis using explicit solver was considered for the large deformation. The details of the FEA model validation are shown in Sect. 1.

In order to ensure that the FEA model is validated, the nodal displacements corresponded to the locations of extensometers are extracted. The maximum failure load and slope of the failure load against the displacement curve within a linear region are calculated for both experimental data and FE model data based on Eqs. (1–2).

$$Load_{failure} = \max(F) \tag{1}$$

$$slope = \lim_{w=R} \frac{F(i+w) - F(i)}{w} \tag{2}$$

where $Load_{failure}$ and $slope$ are maximum failure load and slope of the failure load against a displacement curve within a linear region, respectively; F is the load matrix during the entire experiment and FEA simulation; I the data index; $F(i+w) - F(i)$ is the data length between two data points under the width of w within a linear region.

The difference in percentage between experimental data and FEA model data are calculated:

$$Difference_{load} = \frac{\left|Load_{experimental\_failure} - Load_{FE\_failure}\right|}{Load_{experimental\_failure}} \times 100\% \tag{3}$$

$$Difference_{load} = \frac{\left|slope_{experimental} - slope_{FE}\right|}{slope_{experimental}} \times 100\% \tag{4}$$

where $Difference_{load}$ and $Difference_{load}$ are difference in percentage between experimental data and those from the FEA mode; $Load_{experimental\_failure}$ and $Load_{FE\_failure}$ are failure load of the experiments and FEA model, respectively (calculated based on Eq. (1)); $slope_{experimental}$ and $slope_{FE}$ are slope of the experiments and FEA model, respectively (calculated based on Eq. (2)).

The FEA model will be validated when both maximum failure load and slope are within the defined $threshold_{load}$ and $threshold_{slope}$:

$$FEA\,model\,is\,validared\,when \begin{cases} Difference_{load} > threshold_{load} \\ Difference_{slope} > threshold_{slope} \end{cases} \tag{5}$$

Based on the validated model, a parametric FEA study was carried out to provide a robust database for the intelligent study of the strength of the joints by using different adherend and adhesive material properties. The same geometry as Fig. 1 is employed. Based on the literature, the elastic modulus of the materials used for the adherends and the adhesive has a significant effect on the joint performance. As the cohesive zone model is used for modelling the adhesive, fracture toughness (cohesive energy) in tension and shear are the most key parameters compared to other cohesive parameters.

Therefore, the range of the elastic modulus of the upper and lower adherends is from 60 to 200 GPa with 4 intervals of a 35 GPa gap (60, 95, 130, 165, 200 GPa), which covers a wide variety of materials including aluminium, composites and steel. The range of fracture toughness in tension covers from 0.2 to 1.6 N/mm with 4 intervals of a 0.35 gap (0.2, 0.55, 0.9, 1.25, 1.6 N/mm), and the fracture toughness in shear covers from 0.4 to 8 N/mm with 4 intervals of a 1.9 gap (0.4, 2.3, 4.2,

6.1, 8 N/mm), respectively. Due to the same failure load by exchanging the material properties between upper and lower adherends, in general, a total of 375 models were calculated as a database.

4.3 Machine Learning Model and Optimization Stage

Machine learning model: Numerous deep learning algorithms have been widely used for different manufacturing applications. Each deep learning algorithm has its pros and cons. For instance, long short-term memory (LSTM) is one of the most popularly used deep learning algorithms in various manufacturing applications [22]. However, LSTM is suitable for processing time-series data but failure load prediction model is not trained by using time-series data. DNN has been successfully used for manufacturing fault detection, health monitoring, etc. [23–25]. Therefore, DNN is considered as a suitable deep learning algorithm to build the non-linear relationship between input parameters and adhesive strength. The reason to choose DNN for this application includes the following aspects: (i) the time complexity of DNN is relatively low and robust (Table 2 shows time complexity analysis for DNN [24]); (ii) prior knowledge regarding bonded composite joints is not required to build the model instead of traditional analytical methods; (iii) DNN can be updated when new data are generated so that the knowledge in the DNN can be continuously updated.

Figure 5 shows the structure of the DNN for adhesive strength prediction. Table 3 shows the input and output of the DNN. All the parameters are stored in input vector$\mathrm{X} = [\mathrm{E1}, \mathrm{E2}, \mathrm{G1}, \mathrm{G2}]$. The adhesive strength is the output variable. In order to build non-linear relationship between the input and the output, hidden layer(s) is also introduced. First hidden neurons $\mathrm{Y1} = [\mathrm{Y}_1, \mathrm{Y}_2, \mathrm{Y}_3, \ldots, \mathrm{Y}_m]$ are calculated below:

$$Y1_j = \sum_{i=1}^{4} X_i w_{ij} + b_1 \tag{6}$$

where i, j and k are counters for X, Y and Z (i is from 1 to 4, j is from 1 to m, Z is 1); w and b_1 are weight and bias, respectively (Fig. 5).

Rectified Linear Unit (ReLU) is also introduced to improve non-linearity, so that the non-linear problem relationship between input and output can be sorted [30]. The calculation is as follows:

$$Y_j^{'} = \max(0, Y1_j) \tag{7}$$

The following Hidden layers $Ym_j^{'}$ are calculated based on the results of previous hidden layer until reaching the last hidden layer:

Table 2 Time complexity of machine learning methods

Machine learning method	Time complexity	Factors
DNN/ANN (Ou and Murphey (2007)) [26]	$O([3 \cdot d \cdot H + 4 \cdot O \cdot (H+1) + 3 \cdot H] \cdot N)$	d: dimension of feature space H: number of hidden nodes O: output nodes N: the number of training data examples
Convolutional Neural Network (CNN) (He and Sun (2020)) [27]	$O(D \cdot N \cdot L \cdot S^2 \cdot M^2 \cdot e)$	L: number of input variables N: number of filters (width) S: spatial size (length) of the filter M: size of the output D: number of convolutional layers (depth) e: number of epochs
Long short-term memory (LSTM (Sak et al. (2014)) [28]	$O(n_c \cdot n_c \cdot 4 + n_i \cdot n_c \cdot 4 + n_c \cdot n_o \cdot + n_c \cdot 3$	n_c: The number of memory cells n_i: the number of input units n_o: the number of output units
Support vector machine (SVM) (Fei et al. (2019)) [29]	$O(N^3)$	N: vectors
Multilayer perceptron	$O(n \cdot M \cdot P \cdot N \cdot e)$	n: input variables M: number hidden neurons P: number outputs N: number of observations e: Number of epochs

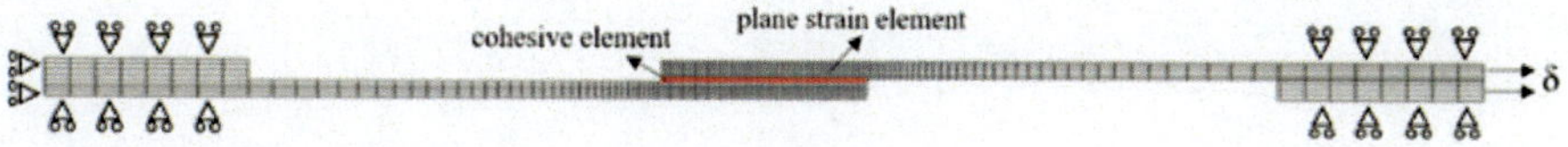

Fig. 4 Load and boundary conditions of the cohesive zone model

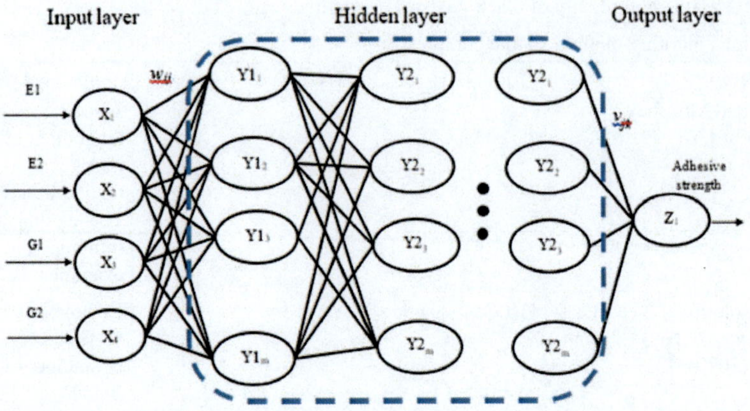

Fig. 5 Design of adhesive strength prediction DNN

Table 3 The input and output of the DNN

Input vector (X)	Output variable (Y)
E1 (elastic modulus of the top adherends)	Adhesive failure strength
E2 (elastic modulus of the lower adherends)	
G1 (mode I fracture toughness of adhesive)	
G2 (mode II fracture toughness of adhesive)	

Table 4 Difference between experiment and FE model data

	Specimen 1	Specimen 2	Specimen 3	FEA prediction
Failure load (N)	8981.625	9488.752652	9957.480266	9432.459126
Failure load difference	5.020%	0.593%	5.273%	
Slope	30,201.21	31,972.53	31,822.0348	32,960.92238
Slope difference	9.138%	3.0913%	3.5790%	

$$Ym_j = \sum_{i=1}^{m} Y_j{}' w_{ij} + b_m \tag{8}$$

$$Ym_j{}' = \max(0, Ym_j) \tag{9}$$

Table 5 Training accuracy benchmark

	Average training accuracy (MSE)	Average Testing accuracy (MSE)	Worst training accuracy (MSE)	Training time (s)	Prediction time (s)
DNN	0.0615	0.0744	0.0742	5.87	1.38
ANN	0.0739	0.0983	0.086	3.39	0.86
CNN	0.180	0.2258	0.202	5.11	1.23
LSTM	0.171	0.2320	0.18	16.41	2.11
RNN	0.167	0.1968	0.167	5.36	1.56

Table 6 Optimization results benchmark

	Improved FFO	**PSO**	**NSGA-II**	**GA**	**SA**
Iterations to achieve optimal results	9	20	23	23	24
Optimized result (N)	19,044.65	19,044.65	19,044.65	19,044.65	19,044.65

where $Ym_j^{'}$ is non-linear hidden neurons with ReLU. Output Z_k can be calculated based on hidden layer(s) $Y_j^{'}$ as below:

$$Z_k = \sum_{j=1}^{m} Ym_j v_{jk} + b \tag{10}$$

where is w and b_1 are weight and bias, respectively.

To train the DNN, collected data from FEA analysis will be utilized. In order to train/re-train the DNN model, the loss function with Mean Square Error (MSE) is defined in Eq. (11):

$$MSE = \frac{\sum_{l=1}^{a} (Z_l - yZ_l)^2}{a} \tag{11}$$

where a is the total quantity of output, y_l is ground-truth for the l-th output, yp_l is l-th predicted output.

The loss function can be minimized to train the DNN by using an improved gradient descent optimiser (improved in Step 4 with a decaying learning rate). The optimization process is explained as follows:

1. Define the weight $W_{DNN} = [w_{ij}, v_{jk}]$ and bias $b_{DNN} = [b_1, b_2]$ of the DNN; initialize the W_{LSTM} and b_{LSTM} with random values;
2. Define the maximum epoch time T;

3. Optimize the value of W_{DNN}, b_{DNN} by minimising the loss function with gradient descent:

$$W_{DNN}{}^{t+1} = W_{DNN}{}^{t} - \varepsilon(\frac{\partial Loss}{\partial W_{ANN}}) \tag{12}$$

$$b_{DNN}{}^{t+1} = b_{DNN}{}^{t} - \varepsilon(\frac{\partial Loss}{\partial b_{DNN}} \tag{13}$$

   where t is the current epoch time; ε is the learning rate. Traditionally, the learning rate is a constant value.
4. In order to improve gradient descent optimization, the learning rate ε decays over training epochs to approach optimal parameters easily:

$$\varepsilon = \varepsilon_0/(1 + k \cdot t) \tag{14}$$

   where k is the hyper-parameter.
5. Repeat the above the steps 3 and 4 until reaching the maximum epoch time.

Fruit Fly optimization: Based on the trained DNN, input parameters can be optimized by a designed improved Fruit Fly Optimization (FFO) algorithm to achieve optimal adhesive strength. The optimization fitness is adhesive strength.

$$fitness = \min(Z) \tag{15}$$

FFO is a swarm algorithm which is initially inspired by foraging behaviors of the fruit fly, smell and vision are the main functionalities of searching food [31]. A fruit fly swarm approaches food source by using osphresis organ to detect food sources around other flies within its swarm. When the fruit fly swarm is close to the food source, fruit flies will use a vision system to approach food source further [32]. The FFO algorithm mimics the foraging behavior to achieve the optimization purpose. The main advantages of FFO are: (i) FFO can search food with multiple fruit flies in parallel, so that the algorithm can achieve convergence quicker with robust result; (ii) FFO has a good capability to avoid local optima; (iii) the mechanism of FFO is simple to develop. In more detail, FFO consists of six main steps:

1. Initialization: Generate a large initial population of M fruit flies randomly. The generated fruit flies are selected to be swarm centers. Each fruit fly contains 1 set of adhesive joint parameters, including E1, E2, G1 and G2. One example of a fruit fly is illustrated below with labels. A fruit fly is generated as below:

$$\text{Fly}_v = \text{LB}_v + (UB_v - \text{LB}_v) \times rand() \tag{16}$$

where v(v = 1, 2, 3, 4) is for a fly, UB_v and LB_v are the upper boundary and lower boundary for the v-th parameter, respectively. $rand()$ is a random value between [0,1].

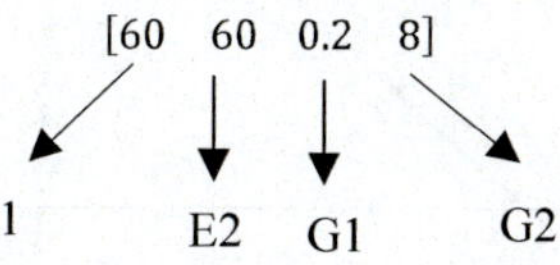

2. Swarm centers selection: Predict adhesive strength with the trained DNN and calculate fitness of each fruit fly based on Eq. (16); Select m (M > m) fruit flies with the best fitness.
3. Determine the maximum iteration (I_{max}) for the entire optimization calculation.
4. Smell-based search: Create 1 sub-population of n fruit flies near each swarm center; Change all the parameters as follows (an example of smell-based search is illustrated below):

$$\text{Fly\_sub}_v = \text{Fly}_v \times (1 + a \times (rand() - 0.5)) \tag{17}$$

where Fly_sub is the randomly generated sub-population of fruit flies. a is the searching step.

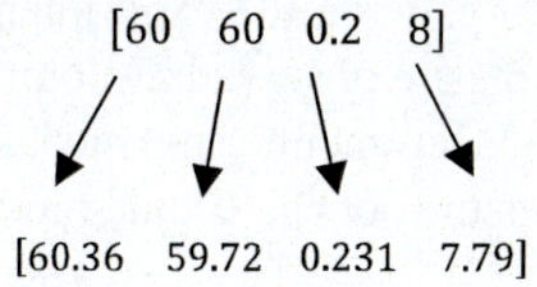

5. Vision-based search: Calculate $fitness$ of all the sub-population of fruit flies. If the $fitness$ is better than that of the swarm center, the fruit fly with better $fitness$ will replace the swarm center. For the purpose of avoiding a local optima, there is a possibility of accepting a fruit fly with the worse result to achieve global optima if Eq. (18) is accepted:

$$c > rand() \tag{18}$$

$$c = \exp(-\left|fitnes_{sub} - fitnes_{centre}\right|/I) \tag{19}$$

where c is a coefficient to decide whether the new worse result can be accepted or not; I is the counter of the current iteration.
6. Repeat Steps 4 and 5 until reaching the maximum iterations I_{max}.

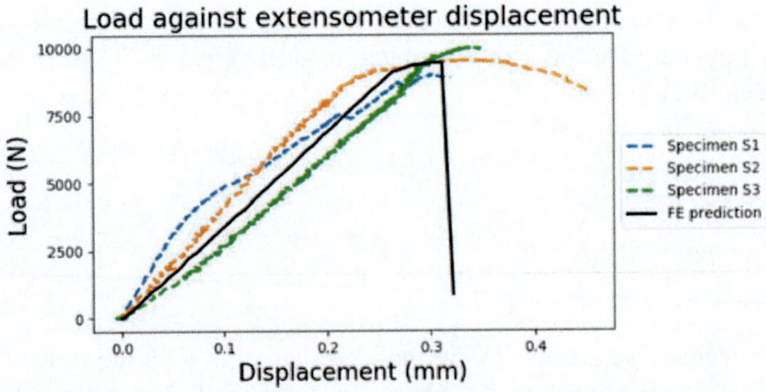

Fig. 6 Comparison of load vs. extensometer displacement relations from experiments and FEA prediction

5 System Development and Analysis

5.1 Experiment and FEM Validation

The validation of a static models is performed by comparing load vs. extension relations with experimental results, as shown in Fig. 6. Table 4 shows the difference according to Eqs. (3–4), and the $threshold_{load}$ and $threshold_{slope}$ are 10% and 10%, respectively. The nodal displacements taken from FE models are corresponded to the positions of the reflective pads on the experimental specimens. A good agreement on the stiffness of the joints are observed by comparing the results from experiments and FEA, although the measurements from the laser extensometer is fluctuating. The comparison also validates the cohesive zone modelling technique.

5.2 Machine Learning Model

For the developed DNN model, different machine learning algorithms are compared for benchmark. To ensure the validation of model accuracy, among 375 data samples, 80% (300 samples) were used for training, about 10% (38 samples) were used for validation and about 10% (37 samples) were used for testing. Figure 7 shows the benchmark for both training accuracy and testing accuracy. Table 5 shows Training accuracy benchmark. In order to show robustness of each algorithm, all the algorithms have been run for 10 time. According to the comparison, it can be seen that the average training accuracy of the DNN is 20.16%, 192.68%, 178.05%, 171.54% better than that of ANN, CNN, LSTM and RNN, respectively. The training and prediction time are 5.87 s and 1.38 s, respectively. The simulation time of FEA is approximately 300 s for each simulation based on parameter setting described in Sect. 2. It can be

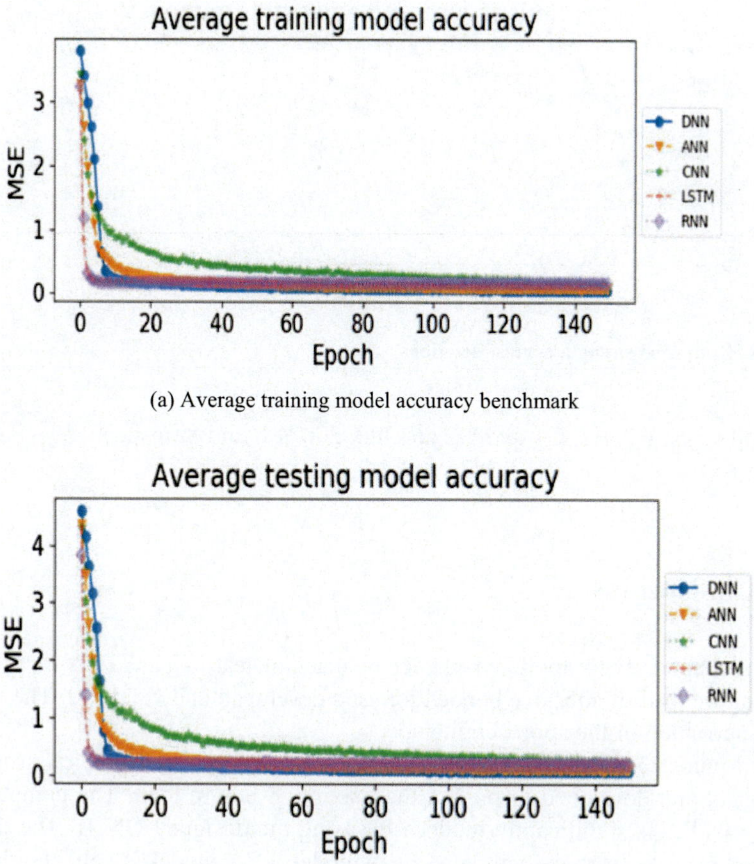

(a) Average training model accuracy benchmark

(b) Average testing model accuracy benchmark

Fig. 7 Training accuracy

seen that the prediction time can be saved by 99.54% by the DNN in comparison with the FEA simulation.

5.3 Optimization

To show optimization efficiency and robustness of the FFO, different optimization algorithms, including PSO, NSGA-II, GA and SA, were compared for benchmark. Each algorithm ran for 10 times, and the average results were taken. It took 9 iterations to achieve optimal results, which is quicker than the convergence time for all the other algorithms. All of the optimization algorithm reached the same optimized result

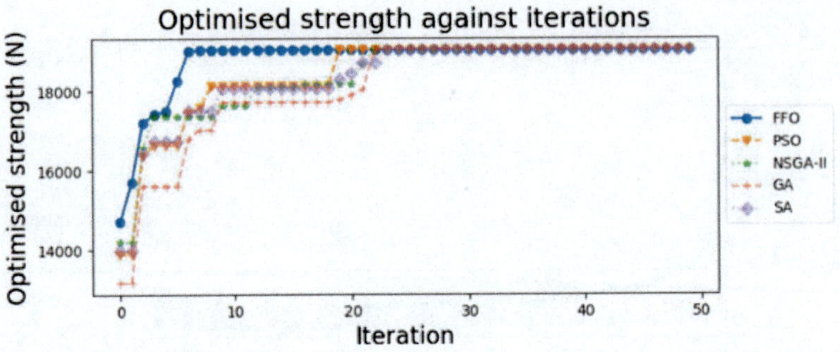

Fig. 8 Optimized strength against iterations

19,044.65 N. E1, E2, G1 and G2 are 200, 200, 1.6 and 8, respectively (Fig. 8 and Table 6).

6 Conclusions

In this chapter, a new approach enabled by machine learning and FEA for predicting the failure load of adhesive bonded joints is developed and validated. The function and innovation of the approach include:

An innovative hybrid machine learning and FEA enabled prediction model is designed and developed to predict the maximum failure load. The computational time of FEA is significantly reduced by using the designed DNNs. The DNNs is trained by data from the validated FEA model. FEA model is validated by using experiments with an Instron 5500R machine. The failure load prediction time is reduced by 99.54%.

- DNNs is designed to prediction failure load, and the algorithm is compared with different machine learning algorithms such as CNNs, LSTM for benchmark. The structure and parameters of the DNNs are justified in order to ensure the best accuracy and efficiency.
- The optimal material parameters of adhesive bonded joints can be found by an innovative optimization algorithm, i.e., FFO, based on the training DNNs model. The selection of FFO is benchmarked, and the optimal parameters can be found within 9 iterations.

However, there are still areas to be further investigated. For instance, there are more parameters generating impacts on the failure load such as treatment method, effect of type, density of fibers, etc. Therefore, more parameters will be investigated and more data will be collected to build a comprehensive DNN model for failure load prediction.

References

1. Cheng F, Hu Y, Lv Z, Chen G, Yuan B, Hu X, Huang Z (2020) Directing helical CNT into chemically-etched micro-channels on aluminium substrate for strong adhesive bonding with carbon fiber composites. Compos Part A: Appl Sci Manufact 135:105952
2. Shi J, Cao W, Wu Z (2019) Effect of adhesive properties on the bond behaviour of externally bonded FRP-to-concrete joints. Compos B Eng 177:107365
3. Capuano G, Rimoli J (2019) Smart finite elements: A novel machine learning application. Comput Methods Appl Mech Eng 345:363–381
4. Ojalvo I, Eidinoff H (1978) Bond Thickness Effects upon Stresses in Single-Lap Adhesive Joints. AIAA J 16(3):204–211
5. Carbas R, da Silva L, Madureira M, Critchlow G (2014) Modelling of functionally graded adhesive joints. J Adhesion 90(8):698–716
6. Stein N, Mardani H, Becker W (2016) An efficient analysis model for functionally graded adhesive single lap joints. Int J Adhes Adhes 70:117–125
7. Stein N, Weißgraeber P, Becker W (2016) Stress solution for functionally graded adhesive joints. Int J Solids Struct 97–98:300–311
8. Bahrami B, Ayatollahi M, Beigrezaee M, da Silva L (2019) Strength improvement in single lap adhesive joints by notching the adherends. Int J Adhes Adhes 95:102401
9. Sadeghi M, Gabener A, Zimmermann J, Saravana K, Weiland J, Reisgen U, Schroeder K (2020) Failure load prediction of adhesively bonded single lap joints by using various FEM XE "FEM" techniques. Int J Adhes Adhes 97:102493
10. Saleh M, Saeedifar M, Zarouchas D, De Freitas S (2020) Stress analysis of double-lap bi-material joints bonded with thick adhesive. Int J Adhes Adhes 97:102480
11. Ramalho L, Campilho R, Belinha J (2020) Single lap joint strength prediction using the radial point interpolation method and the critical longitudinal strain criterion. Eng Anal Boundary Elem 113:268–276
12. Chou J, Tsai C, Pham A, Lu Y (2014) Machine learning in concrete strength simulations: Multi-nation data analytics. Constr Build Mater 73:771–780
13. Zhu J, Zhang W (2018) Probabilistic fatigue damage assessment of coastal slender bridges under coupled dynamic loads. Eng Struct 166:274–285
14. Saadallah A, Finkeldey F, Morik K, Wiederkehr P (2018) Stability prediction in milling processes using a simulation-based Machine Learning approach. Procedia CIRP 72:1493–1498
15. Al-Shamiri A, Kim J, Yuan T, Yoon Y (2019) Modeling the compressive strength of high-strength concrete: An extreme learning approach. Constr Build Mater 208:204–219
16. Putra G, Kitamura M, Takezawa A (2019) Structural optimization of stiffener layout for stiffened plate using hybrid GA XE "Genetic Algorithm (GA)" . Int J Naval Architecture Ocean Eng 11(2):809–818
17. Wu C, Fang J, Li Q (2019) Multi-material topology optimization for thermal buckling criteria. Comput Methods Appl Mech Eng 346:1136–1155
18. Chegini S, Bagheri A, Najafi F (2018) PSO XE "PSO" SCALF: A new hybrid PSO based on Sine Cosine Algorithm and Levy flight for solving optimization problems. Appl Soft Comput 73:697–726
19. Tsiptsis I, Liimatainen L, Kotnik T, Niiranen J (2019) Structural optimization employing isogeometric tools in Particle Swarm Optimizer. J Build Eng 24:100761
20. Huntsman (2015) Araldite 2015 technical data sheet. https://krayden.com/technical-data-sheet/huntsman-araldite-2015-tds. Accessed 21 Oct 2020
21. Sika Company (2009) Sika Primer-204 N. https://gbr.liquidplastics.sika.com/en/sika-liquid-plastics/liquid-roof-waterproofing-accessories/primers/sika-primer-204-n.html. Accessed 21 Oct 2020
22. Campilho R, Banea M, Neto J, da Silva L (2013) Modelling adhesive joints with cohesive zone models: effect of the cohesive law shape of the adhesive layer. Int J Adhes Adhes 44:48–56

23. Cabrera D, Guamán A, Zhang S, Cerrada M, Sánchez R, Cevallos J, Long J, Li C (2020) Bayesian approach and time series dimensionality reduction to LSTM XE "LSTM" -based model-building for fault diagnosis of a reciprocating compressor. Neurocomputing 380:51–66
24. Liang Y., Li W., Lu X., Wang S., 2019. Fog computing and convolutional neural network enabled prognosis for machining process optimization. Journal of Manufacturing Systems. 52: 32–42.
25. Mammone N, Ieracitano C, Morabito F (2020) A deep CNN XE "CNN" approach to decode motor preparation of upper limbs from time–frequency maps of EEG signals at source level. Neural Netw 124:357–372
26. Liang YC, Lu X, Li WD, Wang S (2018) Cyber physical system and big data enabled energy efficient machining optimisation. J Clean Product 187:46–62
27. Ou G, Murphey Y (2007) Multi-class pattern classification using neural networks. Pattern Recogn 40(1):4–18
28. He K, Sun J (2015) Convolutional neural networks at constrained time cost. Proceedings of the 2015 IEEE Conference on computer vision and pattern recognition (CVPR)
29. Sak H, Senior A, Beaufays F (2014) Long short-term memory based recurrent neural network architectures for large vocabulary speech recognition. Available at arXiv:1402.1128
30. Fei X, Shah N, Verba N, Chao K, Sanchez-Anguix V, Lewandowski J, James A, Usman Z (2019) CPS data streams analytics based on machine learning for cloud and fog computing: A survey. Fut Gen Comput Syste 90:435–450
31. Yarotsky D (2017) Error bounds for approximations with deep ReLU networks. Neural Netw 94:103–114
32. Pan W (2012) A new fruit fly optimization algorithm: Taking the financial distress model as an example. Knowl-Based Syst 26:69–74

Enhancing Ant Colony Optimization by Adaptive Gradient Descent

Y. Zhou, W. D. Li, X. Wang, and Q. Qiu

1 Introduction

Ant Colony Optimization (ACO), which is one of the most popular swarm intelligence algorithms [1, 2], has been widely researched for various applications [3–8]. To achieve better performance, many research works have been developing to further improve the ACO algorithm [9].

The first standard ACO algorithm is the Ant System (AS) [3], which has established the basic framework of ACO. It contains two fundamental stages: solution construction and pheromone update. Based on initially constructed solutions, all ants deposit and continuously update pheromones on their travelled paths to achieve optimization. The AS algorithm has been proved to be able to obtain good results in the Travelling Salesman Problems (TSP) and Quadratic Assignment Problems (QAP), though not very competitive yet. In 2000, from a theoretical point of view, Stutzle and Dorigo developed a convergence proof to reveal that its performance highly depends on the pheromone update stage [10]. Thus, two classic variants of AS, i.e., Ant Colony System (ACS) [11] and Max–min Ant System (MMAS) [12], were designed to improve the performance of AS. ACS adopts a random greedy strategy, which selects the next node with the most top heuristic information with a high probability, to conduct intensive search during tour construction. To alleviate weaknesses of AS, such as a slow convergent speed and easy trap into local optima, MMAS applies two strategies: (i) at the end of each iteration, only the best solution is

Y. Zhou · X. Wang · Q. Qiu
School of Computer Science, Wuhan University of Science and Technology, Wuhan, China

W. D. Li (✉)
School of Logistics Engineering, Wuhan University of Technology, Wuhan, China
e-mail: weidong.li@coventry.ac.uk

Y. Zhou · W. D. Li
Faculty of Engineering, Environment and Computing, Coventry University, Coventry, UK

W. Li et al. (eds.), *Data Driven Smart Manufacturing Technologies and Applications*, Springer Series in Advanced Manufacturing,
https://doi.org/10.1007/978-3-030-66849-5_9

used to update the pheromone trail, leading to a better exploitation of good solutions, (ii) the pheromone value for each path is bounded within a range, thus avoiding early maturity.

Overall, the convergence speed and overall solution quality of ACO could be improved by introducing new strategies on the tour construction and the pheromone update, just like the experience provided by ACS and MMAS. These variant algorithms of ACO, however, are not adaptive yet. For a specified problem, choosing appropriate hyper-parameters still requires knowledge and experience. To address this issue, some new parameter optimization schemes for ACO were designed [13]. For instance, some schemes modify parameter setting on a per-instance basis by statistics. Other schemes adjust parameters on the fly by defining strategies like a trade-off between exploration and exploitation of the searching in the solution space. However, the improvements by these themes are instable, instead, leading to even worse performances in some cases [14]. It is worth exploring parameter optimization schemes that have been effectively used in other intelligent algorithms to be integrated into ACO for performance improvement.

Recently, deep learning (DL) is a fast developed branch of artificial intelligence [15]. Most significantly, how to effectively optimize parameters in DL has received a lot of attentions. Gradient descent is one of the most common approaches to tackle this complex problem. The relevant implementation is contained in nearly every DL library (e.g., PyTorch [16], Caffe [17] and Keras [18]). Due to the similar nature in parameter optimization between DL and swarm intelligence algorithms, effective strategies from either side have been considered for borrowing to achieve mutual improvement. For instance, some works of swarm intelligence tried to utilize DL to learn heuristics automatically, which could solve combinatorial optimization problems (COPs) [19–22]. Reversely, some works of DL attempted improving gradient descent by incorporating optimization strategies from swarm intelligence algorithms [23–26]. These studies have motivated us to explore the fusion of DL into ACO somehow for performance improvement. However, current DL approaches for tackling COPs mainly focus on the automatic learning of heuristics, and the performance is less competitive and the scale of COPs to solve based on neural networks is limited. Therefore, in this research, an adaptive Stochastic Gradient Descent (SGD) approach from DL is taken and integrated into ACO for solving hard COPs. To the best of our knowledge, it is still rare to use strategies of DL such as SGD for improving swarm intelligence algorithms in solving COPs. This research aims to find the answer on how to conjunct these two means efficiently and effectively.

With this aim, in this research, an innovative ACO algorithm based on adaptive gradient descent (ADACO) is developed. The major contribution is as follows.

- An ACO algorithm for TSP in the framework of stochastic gradient descent is modeled. We treat the pheromone matrix as the parameters for learning from the random constructed samples and give a definition of gradient for ACO solving TSP. The pheromone deposition and evaporation process of ACO are both explainable in SGD. It lays the basis of integrating adaptive gradient approach into ACO as ADACO.

- Based on the theoretic model, ADACO is designed with random initialization of the pheromone matrix and an adaptive SGD scheme supported with Max–min mechanism. Two key advantages, i.e., asymptotical convergence in ACO and adaptive learning in SGD, are combined to achieve a better solution quality and stability.
- The algorithm was implemented with a parallel scheme. Based on that, the time-complexity of the algorithm is kept similarly as the classic ACO.

The ADACO algorithm was trialed on various sizes of TSP instances in comparison with ACS, MMAS and Iterated Local Search (ILS). The result shows that the ADACO algorithm achieved better accuracy and adaptability.

The rest of this chapter is organized as follows. Section 2 gives a general background of the ACO algorithm and parameter optimization in DL. In Sect. 3, the ADACO algorithm is presented in detail. The experimental and benchmarking results are given in Sect. 4. Section 5 are related works. Conclusion is made in Sect. 6.

2 Background

2.1 Ant Colony Optimization for TSP

TSP is a typical NP-hard problem in combinatorial optimization, which has been popularly used in algorithm validation in theoretical computer science and operations research. In this research, TSP is used to evaluate the effectiveness of the developed ADACO algorithm. In the following, TSP is explained, followed by the modeling process of the ACO algorithm, which lays the base to develop the ADACO algorithm.

The process of TSP is below. Given a complete undirected acyclic graph $G = (V, D)$ with N nodes, where $V = \{v_1, v_2, \cdots, v_N\}$ is a set of nodes and $D = [d_{ij}] \in \mathrm{R}^{N\times N}$ is the weight of an edge $< v_i, v_j >$. The solution of TSP is ordered combinations of all nodes $S = \{x_1, x_2, \cdots, x_N\}, x_i 6 = x_j, i\ 6 = j$, in which $x_i \in \{1, 2, \cdots, N\}$ is indexes of the nodes. The goal of TSP is to minimize the total weights of solutions as shown in Eq. 1.

$$\min L(s) = d(x_N, x_1) + \sum_{i=1}^{N-1} d(x_i, x_{i+1}) \tag{1}$$

where $L(\cdot)$ is the fitness function and $s \in S$ is a solution.

The first version of the ACO algorithm, i.e., AS, was inspired by the foraging behavior of ants. A group of ants work cooperatively to identify good solutions by a mechanism named stigmergy. That is, the ants communicate through pheromone trails, which are chemical substances emitted by the ants. For the TSP problem, AS is composed of two major stages, i.e., tour construction and pheromone update. In the beginning, the ants travel on the graph to create solutions directed by pheromone trails (tour construction). Since better routes cost less effort, they are more likely to aggregate higher quantities of pheromone deposited by the ants. This is the first procedure of the pheromone update. Thus, the routes with more pheromone left would guide later engaging ants to travel through. It is recognized as a positive feedback process. Besides, a sub-procedure of the pheromone update, which is named evaporation, simulates the chemical matter to become weaker in the air as time goes by.

In the tour construction stage, a group of ants create solutions independently and in a stochastic and step-wise fashion according to a pheromone matrix θ. The probability selection for a next travelled node is below:

$$\Pr\left(x|\theta, s^k\right) = \begin{cases} 0 \ \text{ if } x \in s^k, \\ \frac{\theta(s_k, x)}{\sum\limits_{y \notin sk} \theta(s_k, y)} \quad \text{otherwise} \end{cases} \tag{2}$$

where $\theta = [\theta_{ij}] \in \mathrm{R}^{N\times N}$ presents pheromone content on each edge shared among the ants; $s^k = < s_1, s_2, \cdots, s_k >$ is a temporary sequence of travelled nodes; k is the count for constructing steps; $s^0 = \{\emptyset\}$, and $s^N \in S$; an edge $< v_i, v_j > \in s^k$ is true, if and only if $s_i = v_i \& s_{i+1} = v_j$ exists in s^k.

After the ants finish their tours, the fitness values of the solutions are evaluated. Two special types of ants are identified, i.e., an iteration best ant and a global best ant. The former is the best ant in an iteration, and the latter is the best one among the whole iterations.

In the pheromone update stage, the matrix θ is functioned in receiving feedback from ants. One or more solutions that are created by the ants are used to perform a small amount of deposit of pheromone on paths. In AS, all solutions are employed for pheromone depositing, while in ACS and MMAS only the iteration best ant or the global best ant are chosen for pheromone depositing. There are two conditions related to each edge, i.e., evaporation-only and evaporation-with-update. On the first condition, the edges that are not included in selected solutions will perform an evaporation operation. On the second condition, an additional positive value will be added to the edge. The equation of pheromone update for an edge hvi,v_ji according to a selected solution s is as follow.

$$\theta_{t+1}(i, j) = \begin{cases} (1-\rho)\theta_t(i, j) \ \text{ if } \left\langle v_i, v_j \right\rangle > \notin s, \\ (1-\rho)\theta_t(i, j) + \frac{Q}{L(s)} \quad \text{otherwise,} \end{cases} \tag{3}$$

where $L(\cdot)$ is the fitness function, Q is a constant factor, t is the iteration number, and $\rho \in (0,1)$ is an evaporation ratio.

Based on the above equations, the family of the ACO algorithms could be generalized into a framework as presented in Algorithm 1. Note that the local search module (line 4) is applied to exploit neighbors of a solution, which is a technique to speed up convergence and is not a mandatory procedure.

Algorithm 1 ACO framework pseudo-code

```
1: initialize_ants_data();
2: repeat
3:     solution_construction();
4:     local_search(); // Optional
5:     pheromone_update();
6: until termination_criterion()
7: end
```

2.2 *Stochastic Gradient Descent in DL*

As there are hundreds of millions of parameters in complex DL algorithms, the efficiency of parameter optimization is highly concerned. Furthermore, the loss functions of DL algorithms are always non-convex, making it arduous to obtain a global-optimum set of network parameters. SGD (stochastic gradient descent) algorithms have been increasingly popular to optimize DL algorithms [15, 27].

The basic gradient descent algorithm, a.k.a batch gradient descent, is defined as follow. The objective of parameter optimization in DL algorithms is to minimize a loss function $L(\theta,x)$, where θ is the network parameters, and x is the sample data for learning. During the optimization process, x is full samples of a dataset, so that the gradient is defined as $g(\theta) = \nabla_\theta L(\theta) = \frac{\partial L(\theta,x)}{\partial \theta}$. The update equation of batch gradient descent is below:

$$\theta_{t+1} = \theta_t - \rho \cdot g(\theta_t, x) \quad (4)$$

where ρ is the learning rate, and t is a discrete time-step.

The problem of the batch gradient descent is that a considerable amount of time is needed for calculating an exact gradient based on the entire samples in each iteration. That makes it inapplicable for large datasets. To tackle the issue, a basic SGD was proposed to only select one sample $x^0 \in x$ randomly to approximate gradient in each iteration. It is efficient enough to cope with large datasets. However, the gradient is too rough to cause the objective function fluctuation. Thus, the mini-batch SGD version was further proposed to use multiple samples $x^0 \subset x$ for the update process. It is proved to be more stable than the basic SGD while being still computational efficient.

However, the mini-batch SGD does not guarantee good convergence, as it brings the following issues:

- It is hard to determine a proper learning rate ρ. Too small or too large setting may lead to slow convergence or fluctuation.
- Learning rate decay, such as annealing, is an approach to alleviate problems caused by a static setting for the learning rate ρ. Although it sometimes helps in the loss function converging around the minimum, a pre-defined decay schedule introduces other hyper-parameters. The hyper-parameters are hard to define and not adaptive with characteristics of the training data.
- Besides, all the dimensions of the parameter vector apply the same learning rate ρ. That would ignore discrepancies in frequencies of features in training data. For instance, if training data is sparse, it would prefer to set a larger learning rate for rarely arising features.
- Another challenging problem for treating DL tasks, as loss functions are usually non-convex, is that they are easily trapped in local optima or saddle points.

To overcome these drawbacks and develop applicable solutions, the DL community mainly exploit patterns hidden in each dimension. The basic and mini-batch SGD methods adopt only one learning rate across all dimensions. It fails to utilize unique historical information of each parameter as well as discrepancy among them. Therefore, more intelligent approaches are proposed.

An adaptive SGD could deal with sparse data, which commonly use a square root operation on historical gradients. The infrequent data would receive a larger update, and frequently updated parameters are set to a lower learning rate. A representative algorithm is Adagrad [28], in which the square root of all previous gradients is placed as a denominator of the update equation. However, as the learning rate decreases monotonically, it causes an early decay problem. RMSProp [29] voids the problem introduced by Adagrad. The algorithm substitutes the square root sum to the Root Mean Square (RMS), so that the learning rate not always decreases. Similarly, Adadelta [30] not only adjusts the learning rate by the RMS of the gradient but also considers the RMS of parameter updates. It could alleviate the fluctuation of the learning rate.

Adaptive SGD has a strong potential to overcome the weaknesses of ACO for the following reasons. First, the convergence process of ACO is not adaptive yet, and thereby it is easily trapped in local optima. It could be improved by introducing adaptive SGD methods to jump out of a saddle point or local optima in treating non-convex objective functions. Second, choosing a good parameter setting for ACO usually needs a time-consuming trial-and-error process on different datasets. Based on the theory of adaptive SGD, it could, e.g., randomly initialize the pheromone matrix or design an adaptive evaporation scheme to avoid the above parameter choosing process. Besides, the hidden features of each dataset could be hardly found when historical optimization information is ignored. For instance, the evaporation rate is static during the iteration process, and the amount of pheromone deposition is also the same for the edges in a selected route. If the historical update of the pheromone matrix could be exploited, ACO may adapt to datasets with different characteristics. Based on this analysis, the necessity of fusing ACO and adaptive SGD is clear.

3 Adaptive SGD-Based Ant Colony Optimization

In this section, the integration of adaptive SGD and ACO as an ADACO algorithm is developed in enhancing the classic ACO. First, a loss function that minimizes the expectation of error for ACO to solve TSP is defined. Then, the gradient in the loss function is calculated and an adaptive SGD is designed to solve the problem.

This ADACO algorithm over the classic ACO algorithm focuses on the pheromone update stage. The difference between the ADACO algorithm and the classic ACO algorithm is shown in Fig. 1. In the classic ACO, the pheromone update stage is divided into two steps, i.e., evaporation and deposition. Pheromone on all edges evaporates by the learning rate ρ. Then, the density of pheromone on edges in selecting the elite solution is strengthened. The ADACO algorithm treats the pheromone update as a parameter optimization process: (1) the quantity of pheromone deposited is calculated and saved. It indicates some parts of gradient information for the selected edges are more probably included in the global optimum; (2) an approximate gradient is computed for each edge by utilizing the deposited pheromone value; (3) an adaptive pheromone update scheme is designed by considering the gradient and historical gradient information.

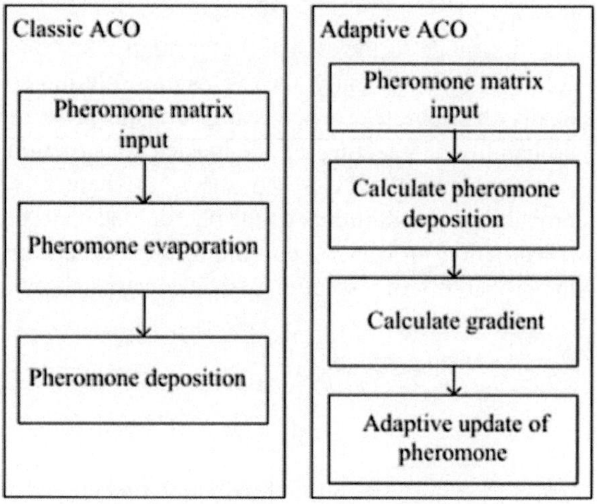

Fig. 1 Comparison of classic ACO and our adaptive ACO in pheromone update stage

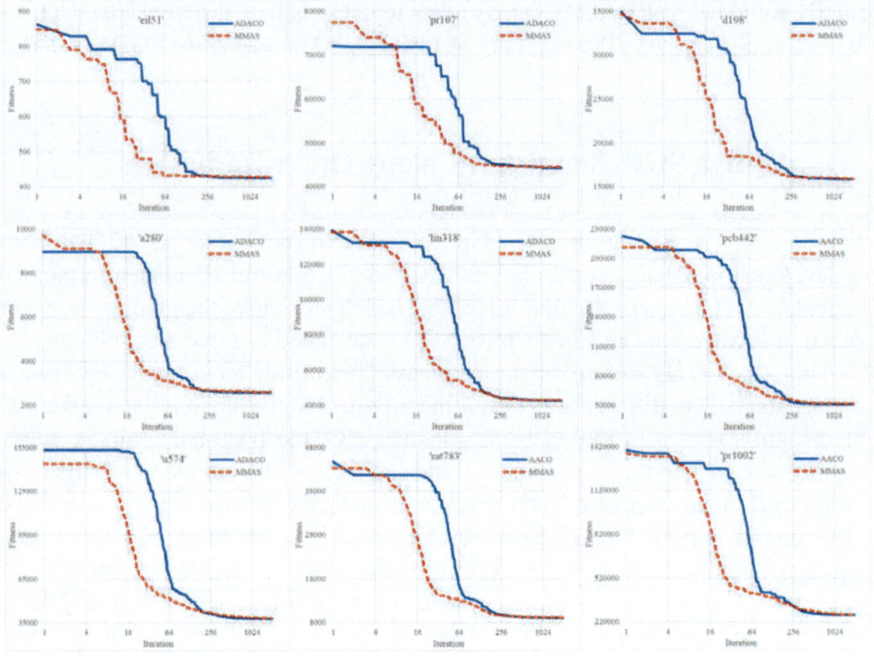

Fig. 2 Convergence curves of ADACO and MMAS (horizontal axis is log scale)

3.1 Loss Function and Gradient Definition for ACO

The first stage of the ADACO algorithm is for tour construction. Since ants create solutions independently, the events of a solution constructed are independent identically distributed. Then looking into the process of constructing a solution, the probability of selecting a node in each construction step can be defined by a conditional probability $P(x_{k+1} = s_{k+1}|\theta, s^{k})$, where k is the step number. The selection probability is conditional on the pheromone matrix θ, since it is static during the tour construction process. According to the theorem on multiplication of probabilities, the probability of a solution can be constructed by the following equation.

$$P(X = s|\theta) = \prod_{k=1}^{N} Pr\left(x = s_k|\theta, s^{k-1}\right)\Bigg) \quad (5)$$

where Pr is the probability distribution function of selecting nodes defined in Eq. (2).

The objective of the ADACO algorithm is to maximize accuracy of randomly constructed solution s following probability distribution P as below:

$$A(\theta) = \frac{L(s^*)}{L(s)} \quad (6)$$

where $L(s^*)$ is the fitness of the global best solution, $L(s)$ represents the fitness of a randomly constructed solution s, and θ is the variable of the function means the goal is to find a optimal pheromone matrix that the accuracy equals to 1. Unlike SGD takes existing samples for learning, the ADACO algorithm generates solutions in the tour construction process. It is stochastic, and therefore the loss function could be defined as minimizing the expectation of error ratio $1 - A(s)$. Since the solutions are randomly sampled, a L2 regularization term is also added to avoid overfitting. The loss function is defined as follow:

$$\begin{aligned} J(\theta) &= \mathrm{E}[1 - A(\theta)] + \frac{1}{2}\ell^2(\theta) \\ &= \mathrm{E}[1] + \mathrm{E}[-A(\theta)] + \frac{1}{2}\ell^2(\theta) \\ &= 1 + \sum_{s \in X^N} -\frac{L(s^*)}{L(s)} \cdot P(X = s|\theta) + \frac{1}{2}\|\theta\|_2^2 \end{aligned} \tag{7}$$

where 1-A(θ) represents the error ratio, ℓ^2 is the L2 norm and its coefficient is 0.5. The goal of ADACO in solving TSP is to minimize the loss function $J(\theta)$. Thus, the gradient of the loss function is deduced as follow:

$$\begin{aligned} \nabla J(\theta) &= \frac{\partial J}{\partial \theta} = \left[\frac{\partial J}{\partial \theta_{ij}}\right]_{N \times N} \\ &= \left[\frac{\partial\left(\sum_{s \in X^N} -\frac{P(X=s|\theta)}{L(s)} + \frac{1}{2}\theta_{ij}^2\right)}{\partial \theta_{ij}}\right] \\ &= \left[-\sum_{s \in X^N} \frac{1}{L(s)} \cdot \frac{\partial P(X = s|\theta)}{\partial \theta_{ij}} + \theta_{ij}\right]_{N \times N} \end{aligned} \tag{8}$$

where $L(s^*)$ is not prior known and fixed for the instance of the TSP problem. Thus, it is set to equal to 1 for simplicity in Eq. (8). Then, according to Eq. (5), the partial derivative of construction probability P is defined by chain rules to get the following equation:

$$\begin{aligned} \frac{\partial P(X = s|\theta)}{\partial \theta_{ij}} &= \frac{\partial}{\partial \theta_{ij}}\left[e^{\ln P(X=s|\theta)}\right] \\ &= P(X = s|\theta) \cdot \frac{\partial}{\partial \theta_{ij}}[\ln P(X = s|\theta)] \end{aligned}$$

$$= P(X = s|\theta) \cdot \frac{\partial\left[\sum_{k=1}^{N} \ln Pr\left(x = s_k|\theta, s^{k-1}\right)\right]}{\partial\theta_{ij}}$$

$$= P(X = s|\theta) \cdot \sum_{k=1}^{N} \frac{\partial \ln Pr\left(x = s_k|\theta, s^{k-1}\right)}{\partial\theta_{ij}} \tag{9}$$

Let $T_{ij}(\theta), \frac{\partial \ln P(X=s|\theta)}{\partial\theta_{ij}} \sum_{k=1}^{N} \frac{\partial\left[\ln Pr\left(x=s_k|\theta,s^{k-1}\right)\right]}{\partial\theta_{ij}}$, based on Eqs. (8) and (9), the following equation can be obtained:

$$\nabla J(\theta) = \left[-\sum_{s\in X^N} \frac{1}{L(s)} \cdot \frac{\partial P(X = s|\theta)}{\partial\theta_{ij}} + \theta_{ij}\right]_{N\times N}$$

$$= \left[-\sum_{s\in X^N} \frac{T_{ij}(\theta)}{L(s)} \cdot P(X = s|\theta) + \theta_{ij}\right]_{N\times N}$$

$$= -\mathrm{E}\left[\frac{T(\theta)}{L(s)}|\theta\right] + \theta \tag{10}$$

As a result, the evolution process of the ADACO algorithm could be viewed as a learning process as below. That is, it iteratively updates the learning parameter (same as pheromone) matrix to reduce results of the loss function:

$$\theta := \theta - \rho\nabla_\theta J = \theta - \rho\left(-\mathrm{E}\left[\frac{T(\theta)}{L(s)}|\theta\right] + \theta\right)$$

$$= (1 - \rho)\theta + \rho\mathrm{E}\left[\frac{T(\theta)}{L(s)}|\theta\right] \tag{11}$$

From Eq. (11), it is noted that the L2 regularization term in the update equation is the same as the pheromone evaporation in Eq. (3). This reveals that the evaporation process is actually functioned as weight decays. Then, the expectation item in Eq. (11) that is related to the pheromone deposition process in ADACO is discussed in the next section.

3.2 Calculation of Gradient in ACO

Since it is impossible to list all solutions as samples for learning, a sample of the expectation could be substituted for the gradient in Eq. (11) and yield the following equation:

$$\theta := (1-\rho)\theta + \rho\frac{T(\theta)}{L(s)} \tag{12}$$

In the AS algorithm, all ants deposit pheromone on their travelled paths. The AS algorithm follows the direction of Eq. (12) by averaging multiple samples as $\frac{1}{M}\sum_{i=1}^{M}\frac{T(\theta)}{L(s)}$.

However, the convergence speed is slow due to the samples may contain'bad' solutions. In ACS and MMAS, only the best ants are allowed to deposit pheromone, which is a way to accelerate convergence speed on a random greedy basis with a small loss of accuracy. According to the theory of SGD, learning by random gradient perform better than by exact gradient in cases of non-convex problems [31]. Just like estimating a gradient that is based on a mini-batch from whole samples in the SGD, the solution created by the iteration-best-ant is selected to calculate a proximate gradient.

Next, the calculation process of $T(\theta)$ in ADACO is shown below in detail:

$$\begin{aligned}
T_{ij}(\theta) &= \sum_{k=1}^{N}\frac{\partial}{\partial\theta_{ij}}\left[\ln Pr\left(x = s_{k+1}|\theta, s^k\right)\right] \\
&= \sum_{k=1}^{N}\frac{\partial}{\partial\theta_{ij}}\left[\ln\frac{\theta(s_k, s_{k+1})}{\sum\limits_{y\notin s^k}\theta(s_k, Y)}\right] \\
&= \sum_{k=1}^{N}\frac{\partial}{\partial\theta_{ij}}\left[\ln\theta(s_k, s_{k+1}) - \ln\sum_{y\notin s^k}\theta(s_k, Y)\right] \\
&= \sum_{k=1}^{N}\frac{\partial\ln\theta(s_k, s_{k+1})}{\partial\theta_{ij}} - \frac{\partial\ln\sum\limits_{y\notin s^k}\partial\theta(s_k, Y)}{\partial\theta_{ij}} \\
&= \sum_{i=s_k,k=1}^{N}\frac{1_{\langle i,j\rangle\in s}}{\theta_{ij}} - \frac{1}{\sum\limits_{Y\notin s^k}\theta_{iY}} \\
&= \sum_{i=s_k,k=1}^{N}\frac{1_{\langle i,j\rangle\in s}}{\theta_{ij}} - \frac{\frac{\partial_{ij}}{\sum\limits_{\nu\notin\cdot k}\theta_{ij}}}{\theta_{ij}} \\
&= \sum_{i=s_k,k=1}^{N}\frac{1_{\langle i,j\rangle\in s} - Pr\left(x = j|\theta, s^k\right)}{\theta_{ij}} \\
&= \frac{1_{\langle i,j\rangle\in s} - Pr\left(x = j|\theta, s_k = i, s^k\right)}{\theta_{ij}},
\end{aligned} \tag{13}$$

where $1_{\langle i,j\rangle \in s}$ is an indicator function defined to be 1 if $\langle i,j\rangle \in s$, else 0. Note that the summation is removed in the last step of the above equation. The reason is that the start node i of an edge $\langle i,j\rangle$ is appeared in the s only once, so that the probability of selecting the edge just calculated once. From Eq. (13), Eq. (11) can be written as:

$$\begin{aligned}\theta_{t+1} &= (1-\rho)\theta_t + \rho\theta'_t \\ \theta'_t &= \left[\theta'_{ij}\right]_{N\times N} \\ &= \left[\frac{1_{\langle i,j\rangle \in s} - Pr_\theta(x=j|s_k=i)}{L(s)\cdot\theta_{ij}^{(t)}}\right]_{N\times N}\end{aligned} \tag{14}$$

where Pr_θ $(x = j \mid s_k = i)$ denotes the probability of selecting node j when the ant is in node i. Equation (14) demonstrates that the pheromone is strengthened on the route of the iteration-best-ant solution, and weaken otherwise.

It needs to compute the probability of all the edges if directly using Eq. (14) to calculate the gradient. It also requires treating multiple starting nodes for the solution of the TSP problem that is a circle. Considering the above conditions, it is time-consuming and hard to implement. Therefore, the pheromone update scheme of the classic ACO algorithm in Eq. (3) is reorganized to replace the theory gradient as follow:

$$\begin{aligned}\rho\theta'_{ij} &= 1_{\langle i,j\rangle \in s}\cdot\frac{Q}{L(s_t^*)} \\ \theta'_{ij} &= 1_{\langle i,j\rangle \in s}\cdot\frac{Q}{\rho L(s_t^*)}\end{aligned} \tag{15}$$

where s^*_t is the iteration-best-ant solution in the t-th iteration. It makes sense because it follows the direction that pheromone on the iteration-best-ant solution will strengthen given by Eq. (14). Accord to the Eqs. (11), (12), (14) and (15), the gradient of the ADACO is defined as follow:

$$\begin{aligned}g_t &= \theta_t - \theta' = \left[g_{ij}^{(t)}\right]_{N\times N} \\ g_{ij}^{(t)} &= \theta_{ij}^{(t)} - \theta_{ij}^{(t)'} = \theta_{ij}^{(t)} - 1_{\langle i,j\rangle \in s}\cdot\frac{Q}{\rho L(s_t^*)}\end{aligned} \tag{16}$$

The above process shows that the ADACO includes an instance of SGD, and both the pheromone deposition and pheromone evaporation process are explainable in the SGD framework. The above theoretical models can assure that ADACO can achieve robust convergence based on SGD, which leads to the idea of applying adaptive SGD to enhance ACO.

3.3 Design and Implementation of ADACO

According to the theoretical models, it proves adaptive SGD for ACO is feasible. In this section, how to design such adaptive SDG and integrate it into ADACO is introduced.

As analyzed earlier, the classic ACO applies a static approach to update the pheromone matrix. The learning rate ρ is fixed in the iterations of optimization, as well as θ' is the same on edges which pheromone would enhance and otherwise zero. It fails to exploit update histories of each edge and ignores discrepancies in the edges of a candidate solution for pheromone depositing. Thus, several problems may occur as the following two categories:

(1) Convergent inefficiency:

- The pheromone update for an edge has a high possibility to fluctuate because no previous knowledge for it is considered.
- The candidate solution is surely different from the global optimum for most of the time, while the edges of the solution share only an incremental value. Sparse features, like that an edge could be hardly constructed but is the part of that global optimum, would be ignored.

(2) Hyper-parameter dependency:

- Choosing hyper-parameters such as ρ has a high impact on the solution quality. A higher value would cause premature and lower one would introduce slow convergence. It also depends on the problem characteristic that needs some knowledge to select a suitable value.
- Initializing the pheromone matrix also needs experiences of the problems, as the classic ACO provides a strategy for the TSP. However, it is problem dependent that hardly promotes to solve other problems.

To address the above issues, first, the ADACO algorithm is designed to update the pheromone matrix in a per-dimension basis and adaptively in each dimension. In this research, ADACO integrates Adadelta [30], which has not been applied in the combinatorial optimization domain as far as we know, to facilitate the pheromone update process. Meanwhile, it preserves Max–min pheromone limiting scheme which has an asymptotically convergent guarantee. The equation is defined below:

$$
\begin{aligned}
G_t &= \gamma G_{t-1} + (1-\gamma) g_t \odot g_t \\
H_t &= g_t \frac{\sqrt{\Delta G_t + \varepsilon}}{\sqrt{G_t + \varepsilon}} \\
\theta_t &= maxmin(\theta_{t-1} - H_t) \\
\Delta G_t &= \gamma \Delta G_{t-1} + (1-\gamma) H_t \odot H_t
\end{aligned} \tag{17}
$$

where G_t, which is the exponentially decaying average of the squared gradients with a decay factor γ, is element-wise dot product. It functions to restrain fluctuation of the random gradient. Similarly, $\Delta G_t$is an exponentially decaying average of the squared updates with the decay factor γ. It works to suppress fluctuation of the updates of θ. H_t is a corrected update matrix through an approximation to the Hessian of the pheromone matrix, which is calculated by dividing the RMS of the updates by the RMS of the gradients. The maxmin function is defined as below:

$$\max\min(x) = \begin{cases} \theta_{\max} & x > \theta_{\max} \\ x & \theta_{\min} \le x \le \theta_{\max} \\ \theta_{\min} & x < \theta_{\min} \end{cases} \quad (18)$$

where θ_max and θ_min are the up and down limits of a pheromone value.

Second, since parameter initialization is a key issue in SGD, ADACO could benefit from choosing a proper strategy. The classic ACO initializes the pheromone matrix by θ_ij = θ_max, which is calculated according to a greedy constructed solution and θ_min = θ_max/2 N. Apparently, it depends on the specific problem instance. In this work, ADACO set a fixed θ_max for all the problem instances and randomly initializes the pheromone matrix in a fixed range as below:

$$\theta_{ij} = \theta_{\max} - Kr(\theta_{\max} - \theta_{\min}) \quad (19)$$

where K is a constant that control the distribution range and r is a random value between 0 and 1.

Algorithm 2 The pseudo code of the pheromone deposition kernel.

1: **for each** $i = 0 \to n - 1$ **parallel do**
2: delta_theta[s[i]][s[i+1]] += 1.0 / Fitness(s); {Record deposit value on the edge $\langle s_i, s_{i+1} \rangle$}
3: **end for**
4: **end**

In Algorithm 2, delta theta is a N × N matrix that each element corresponds to an edge, and it is initialized with 0;s is an array of the iteration-best-ant solution.

The second part incorporates gradient calculation in Line 2, which uses the deposition value processed in the first part based on Eq. (16). The dimension of the matrix theta is N × N that is mapped one to one with the pheromone matrix. Other matrices, which support the update of the matrix theta, are also N × N. Since the parameters are updated through each dimension, it could be implemented in parallel. According to Eq. (17) of adaptive pheromone update, a parallel implementation is developed. The pseudo code of the adaptive pheromone update is below.

Algorithm 3 The pseudo code of the adaptive pheromone update kernel.

```
for each i = 0 → n^2 − 1 parallel do
    gt[i] = theta[i] - delta_theta[i] / rho;
    acc[i] = beta1*acc[i] + (1-beta1)*gt[i]*gt[i];
    update = gt[i]*sqrt(d_acc[i] + epsilon) / sqrt(acc[i] + epsilon);
    theta[i]=theta[i] - update;
    d_acc[i] = beta1*d_acc[i] + (1-beta1)*update*update;
    {Apply max-min limitation}
    if(theta[i]>THETA_MAX) theta[i]=THETA_MAX;
    if(theta[i]<THETA_MIN) theta[i]=THETA_MIN;
end for
end
```

Therefore, the advantages of the ADOCO algorithm are below:

- The adaptive tuning of gradient would be helpful to avoid being trapped in local optima, and make the converging process more intelligently.
- Since historical information of parameters is exploited, less hyper-parameter dependency should be achieved.
- The algorithm would be more robust and accurate than the classic ACO.

3.4 Complexity

While the ADOCO algorithm only modifies the pheromone update process, the time complexity is analyzed. Algorithm 2 and Algorithm 3 display that there is only one loop to traverse elements of the pheromone matrix. Therefore, the time complexity of ADOCO is $O(n^2)$, which is the same as the classic ACO. In the research, only several arrays to store gradients (gt[n^2]), accumulated squared gradients (acc[n^2]) and accumulated delta (d_acc[n^2]) in Algorithm 2 is added, so that the space complexity is also maintained. Thus, ADOCO keeps complexity at the same level as the classic ACO. Furthermore, as it is implemented in a parallel environment, the extra computational overhead should be trivial.

4 Performance Evaluation

Experiments were conducted to validate the ADACO algorithm on a PC with an Intel(R) Xeon(R) CPU E5-4607 v2 @ 2.60 GHz. The algorithm was implemented on the parallel ACO platform developed by the authors [40], which was written by C + + and OpenCL.

Table 1 Hyper-parameter configuration. N.A. Stand for'Not Applicable'. N is the number of nodes that is A problem specific configuration

Property	Notation	Classic ACO	ADACO
Nodes number	n	N	N
Ants number	m	N	N
Pheromone factor	α	1	1
heuristic factor	β	2	2
RMS decay	γ	N.A	0.95
Epsilon		N.A	1e-7
Heuristic strength	Q	1	1
Evaporation (Learning) rate	ρ	0.2	N.A
Iterations	i	2000	2000

The experiments took a standard set of benchmark instances from the TSPLIB library [32]. Single-precision float point numbers for the pheromone information and city distances were used in the experiments. On the algorithmic level, to conform to the experimental principles adopted by Stutzle et al. [12], key parameters were configured as follows: $m = n$ (n being the number of cities), $\alpha = 1$, $\beta = 2$, and $\rho = 0.2$. Table 1 summarizes the hyper-parameter configuration.

For the program settings, each TSP instance was tested for 2000 iterations and averaged over 25 independent runs, which is the same as the previous high-performance ACOs [33, 40]. The MMAS algorithm, which was designed by Stuzle¨ in [12], was used as a baseline for benchmarking. The reason that MMAS was chosen to conduct a particular comparison is that it is the closest ACO variant to the ADACO. And the results would be meaningful for the theoretic analysis. To further ensure that experimental results are convincing, solution quality comparisons among the results obtained by the AS, ACS, MMAS and ILS algorithms are also provided.

Statistics were applied on the results by the following methods. First, average, maximum, minimum and standard deviation are calculated from each group of the runs. Second, the solution quality is indicated by accuracy $A(s) = L(s^*)/L(s)$, where s^* is the optimal solution, and s is the averaged output.

4.1 Convergence Analysis

The convergence curves of ADACO and MMAS in 9 TSP instances were compared (illustrated in Fig. 2). The curves selected are the best one among the 25 runs for each TSP instance. The convergent speed of MMAS is higher than ADACO in the first 128 iterations. This could be explained that because the learning rate of MMAS is fixed, it converges fast at early stages. However, this would make a higher possibility to stick in a local optimum. ADACO performs smoother during the whole convergence process. Although in the beginning, it drops slower than MMAS, after 256 iterations, it keeps descending slowly and surpasses MMAS finally (it could be better observed through Table 2 in Section IV.C). ADACO converges adaptively due to the exploitation of history gradient based on the deduced theoretic model in this research, and thus it could achieve better results more probably.

4.2 *Parameter Adaptation*

In this section, the hyper-parameter adaptation performance was evaluated by changing the evaporation rate ρ from 0.2 to 0.8 with a stride 0.2. Note that in ADACO, the learning rate is adaptive, so ρ is only used for calculating the gradient. The results in Fig. 3 display a superior accuracy performed by ADACO in all cases. As MMAS uses a fixed learning rate, it can be observed that this hyper-parameter has a high impact on its performance. The bigger learning rate may result in accelerating convergent speed but harmful for searching the global optimum. The results of ADACO is more stable with the help of correcting gradient in each iteration. Besides, it was found that $\rho = 0.2$ is the best setting for both algorithms, showing that it is not only an empirically good choice for the evaporation rate in MMAS but also beneficial for the adaptive gradient calculation in ADACO.

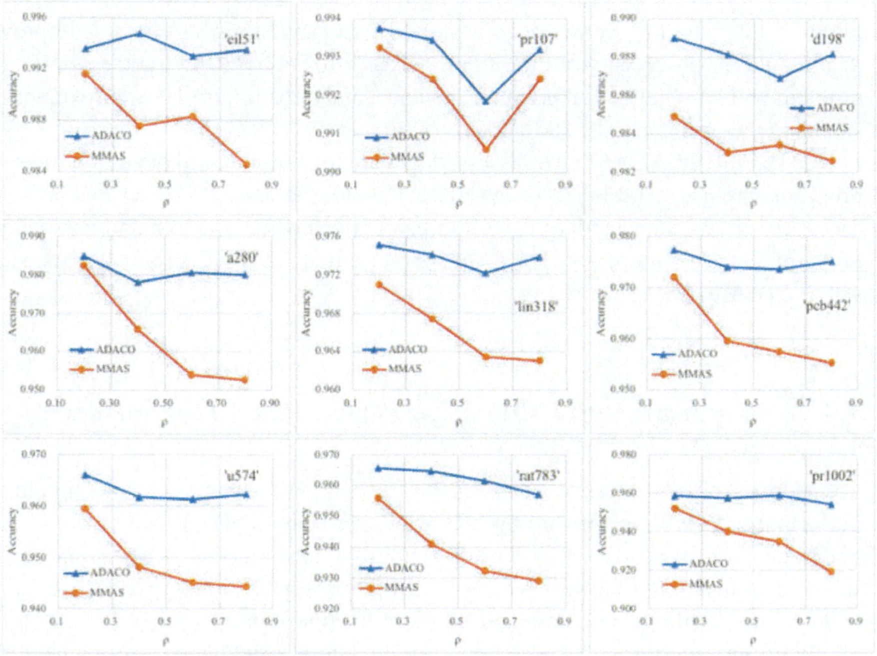

Fig. 3 Results for hyper-parameter adaptation

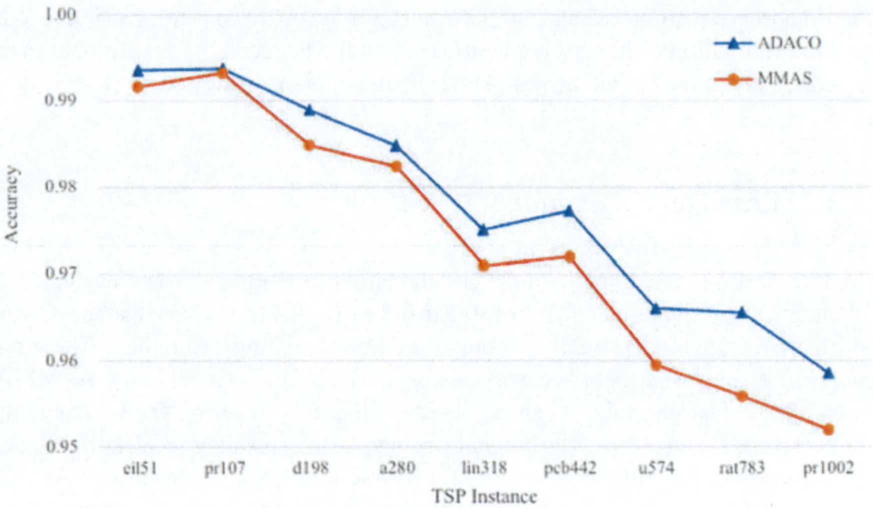

Fig. 4 Average solution quality comparison result

4.3 Solution Quality

Figure 4 illustrates the average accuracy of ADACO and MMAS. A notable improvement could be observed from the figure. Except for'pr107′, the trend is the larger problem will gain better performance advance, and a maximum 1% improvement is shown in the problem with 783 nodes.

The stability of ADACO was assessed by viewing the standard deviation index of program outputs. In Table 2 it shows that ADACO is at least 8% better than MMAS for nearly all cases in standard deviation index; meanwhile, the min and max values of ADACO are also superior. The results indicate that ADOCA is more accurate and stable than MMAS.

4.4 Comparison with Other High Performance TSP Solvers

To ensure the effectiveness of ADACO, the algorithm was compared with other high-performance TSP solvers including ACS with local search, MMAS with local search, and Iterated Local Search (ILS). The original AS algorithm is set as a baseline result. In order to guarantee fairness, each algorithm ran for 25 times in 2000 iterations and average results were calculated. A 2-opt local search operator was chosen for all algorithms ending with'LS' (local search). The percentage deviations from the baseline results in parentheses for all other algorithms are given in Table 3. The result of algorithms without local search illustrates that the ADACO performed better than other variants of ACO due to the effect of adaptive learning mechanism. It was validated that the adaptive SGD indeed increased the accuracy of the ACO as the

Table 2 Stability Comparison with Classic ACOs

Algorithm	Indx	eil51	prl07	dl198	a280	lin380	pcb442	U574	rat783	pr1002
ADACO	Mill	**426**	**44303**	**15845**	**2583**	**42679**	**51193**	**37446**	**9032**	**266,992**
	Mn«	**433**	**44785**	**16129**	**2666**	**43582**	**52456**	**38559**	**9230**	**273351**
	Sid	**1.9209**	**140.9054**	**78.6893**	**23.8459**	**266.859**	**278.2229**	**264.104**	47.2242	**1773.1195**
MMAS	Min	426	44303	15867	2597	42860	51709	37932	9124	268999
	Ha	440	45089	16195	2698	44186	53006	39123	9287	277042
	Sid	3.7639	185.7855	91.6942	24.564	312.8496	350.8479	**316.9181**	**39.4093**	1939.5163
ACS	Min	426	44303	16136	2649	43273	52324	39918	9506	293807
	Max	445	44978	17156	2774	45346	56075	44228	10535	314304
	Sid	4.4956	197.8749	279.103	31.2616	491.102	870.5477	948.5039	296.1372	4101.6553

*The lower the better. Best results are displayed in bold font

Table 3 Comparison with Others High Performance TSP Solvers

Problem	AS	ACS	MMAS	ADACO	1LS	AC5-LS	MMAS-LS	ADACO-LS
eil51	437.36	431.28 (−1.39%)	429.6 (−1.77%)	428.76 (−1.97%)	426.48 (−2.49%)	426.72 (−2.43%)	426.16 (−2.56%)	**426.12 (−2.57%)**
prl07	47257.84	44657.64 (−5.50%)	44604.52 (−5.61%)	44581.56 (−5.66%)	44316.88 (−6.22%)	44333.96 (−6.19%)	44303 (−6.25%)	**44303 (−6.25%)**
d198	17210.6	16713.72 (−2.89%)	16022 (−6.91%)	15955.8 (−7.29%)	15796.84 (−8.21%)	15792.4 (8.24%)	15788.16 (−8.26%)	**15785.96 (−8.28%)**
a280	3046.2	2690.04 (−11.69%)	2625.16 (−13.82%)	2618.72 (−14.03%)	2588.16 −15.04%	2585.52 (−15.12%)	2579 (−15.34%)	**2579 (−15.34%)**
lin318	47091.4	43989.84 (−6.59%)	43286.04 (−8.08%)	43102.64 (−8.47%)	42384.84 (−9.99%)	42261.84 (−10.26%)	42129.16 (−10.54%)	**42116.76 (−10.56%)**
pcb442	61128.44	53847.64 (−11.91%)	52240.6 (−14.54%)	51958.44 (−15.00%)	51222.68 (−16.20%)	51054.68 (−16.48%)	50999.8 (−16.57%)	**50974.16 (−16.61%)**
u574	44758,92	4187132 (−6.45%)	38462.48 (−14.07%)	38202.48 (−14.65%)	37,412.28 (−16.41%)	37212.32 (−16.86%)	37090.96 (−17.13%)	**37022.28 (−17.29%)**
rat783	10934.2	9910.92 (−9,36%)	9212.44 (−15.75%)	9120.04 (−16.59%)	8984.64 (−17.83%)	8879.92 (−18.79%)	**8852.93 (−19.03%)**	**8853.68** (−19.03%)
pr1002	328555.52	304847.68 (−7.22%)	272089.96 (−17.18%)	270248.4 (−17.75%)	263825.92 (−19.70%)	261571 (−20.39%)	261057.52 (−20.54%)	**260806 (−20.62%)**

44242116.76* Deviations from the baseline results are given in parentheses. The low the better. Best results are displayed in bold font theoretic analysis in this research. A local search operator was also appended to check the adaptation of the ADACO, and a positive result was obtained. The ADACO algorithm demonstrated the best performance in 8 TSP instances out of 9. It could be explained that the local search operator helps generate better samples by searching the neighborhood. The ADACO algorithm could learn from these samples adaptively to improve solution quality.

4.5 Time Cost

The time complexity was analysed in the previous section. Figure 5 demonstrates that the computation time for ADOCA and MMAS is nearly the same owing to the parallel implementation of ADOCA. The results are measured with the iteration-best-ant update strategy, which is popular in high-performance ACOs, such as MMAS and ACS. The results demonstrate that ADOCA has a very little extra computational overhead, while achieves much better results.

5 Related Works

There are a lot of works related to the advancement of ACO. They can be found in the latest review [9], which covers studies including categories such as theories, algorithms, and applications, etc. Since this research focuses on the algorithmic aspect, the following related works are reviewed here.

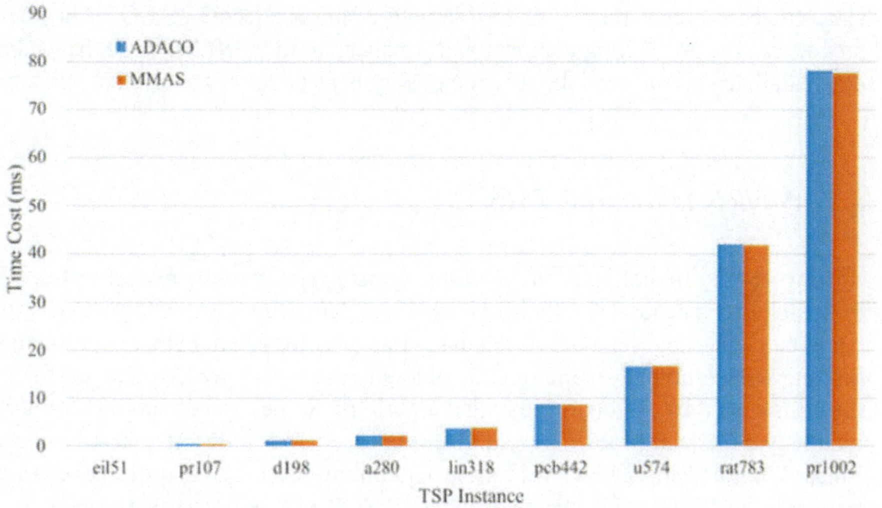

Fig. 5 Average execution time of evaporation kernel in an iteration

5.1 Improving ACO Algorithms

Two key aspects of improving ACOs are accuracy and efficiency. Early works give directions of how to advance the ACO algorithm in the accuracy aspect.

Yang et al. [34] proposed an improved ACO algorithm for solving mobile agent routing problem. A Genetic Algorithm was integrated into ACO to provide a better convergence performance and give a revised global pheromone update principle.

The performance of the algorithm is better for escaping from the trap of a local minimum as well as faster convergence. Elloumi et al. [35] studied the performance of an algorithm hybridizing of ACO and PSO. The junction point is the pheromone deposit by ACO would be transferred to the particle weights of PSO, thus to make the global search more intelligent. Olivas et al. [36] devised a dynamic parameter approach for ACO based on interval type-2 fuzzy systems. The method utilizes fuzzy logic to manipulate the diversity of the solutions, thus to control the exploration and exploitation of ACO. Engin et al. [6] presented an effective hybrid ACO algorithm based on crossover and mutation mechanism for no-wait flow shop scheduling with a criterion to minimize the maximum completion time. The crossover and mutation operators were investigated by introducing a Genetic Algorithm into the ACO algorithm. Chen et al. [37] designed an entropy-based dynamic heterogeneous ACO. The research enhances the stability and precision of classic ACOs with dual heterogeneous colonies and communication by an allotropic mechanism. However, the above research works only tested small-scale problem instances, so their performance could not be guaranteed in large-scale ones. Besides, most works still lack of solid theoretical support for experiments.

Paralleling ACO is also an actively explored research topic especially in the aspect of GPU-based approaches. GPU-based AS [33, 38], ACS [39] and MMAS [33] were proposed with significant progresses in performance. Multi-SIMD CPU parallel was also utilized to accelerate ACO [40] largely against sequential ACO. It has been proved that the ACO computational performance could be effectively enhanced by data-parallelism. However, the solution quality is just kept as the original version.

5.2 Models to Solving TSPs

Solving TSP by modeling is another topic closely related to the research presented in this chapter because the TSP problem is modelled by a probability model that could be tackled by the SGD. Cvetkovic et al. [41] modeled TSP as a problem of discrete semidefinite programming. Semidefinite relaxation was applied to TSP and then it is solved by a branch and bound algorithm. Some preliminary test results of randomly generated problems were given. Focacci et al. [42] integrated an Integer Linear Programming with Constraint Programming (CP). The approach overcomes limitations in dealing with objective functions of CP. TSP and its variant problem were experimented and some satisfactory results were obtained. Xie et al. [43] presented a multi-agent optimization system to perform cooperative search to solve TSP. Memory and rule modules were designed with considering existing knowledge components for TSP. Some positive results were obtained, and they were comparative with several famous TSP solvers. Kool et al. [19] proposed a neural network model based on attentional encoder-decoder to learn heuristics for solving COPs including TSP. It was found to be a feasible approach in achieving results close to traditional TSP solvers. It could save costly development of dedicated heuristics, but at a very high computational price. The TSP instances for testing were limited to 100 nodes. Shi et al. [44] introduced a new method named homotopic convex transformation to smoothen the landscape of TSP. This approach combines the original TSP and a convex-hull TSP by a coefficient to generate a new smoother problem, which is easier to address than original TSP. Experimental results show the algorithm outperformed iterated local search for most TSP cases tested.

The above works have a very strong theoretical background that may inspire future works. Nevertheless, the efficiency is not considered enough that makes the approaches inapplicable in some complex scenarios. The works only experimented with small-scale cases and a considerable amount of computational time was consumed.

6 Conclusions

Swarm intelligent approaches such as ACO have been applied to solve many NP-hard optimization problems. However, the performance of ACO is not satisfactory since the pheromone update is static, result in early maturing and slow convergence. The research in this is chapter is aimed to improve the performance of ACO through adaptive SGD. A theoretic analysis between ACO and SGD is given in the context of solving TSPs, and it can be concluded that ACO is an instance of SGD. Based on this finding, a per-dimensional adaptive gradient descent approach for ACO, namely ADACO, is investigated. In this approach, a form of gradient calculation for ACO to solve TSPs is modelled, and an adaptive parameter update scheme according to the gradient is designed. A broad range of TSP instances varying from 51 to 1,002 nodes were used to evaluate the designed ADACO algorithm. Case studies showed that ADACO achieved improvement over classic MMAS as high as 1% in average accuracy, and meanwhile are more stable. Besides, ADACO is less hyper-parameter dependent due to its contribution to convergent more intelligently. Furthermore, more experiments were conducted to compare the ADACO algorithm with the AS, ACS and ILS to valid its effectiveness. The results show the performance is promising.

More interesting, the proposed algorithm can be generalized. SGD approach potentially can be applied to other metaheuristics including stochastic selection process such as genetic algorithms. ADACO can be also applied to solve other COPs owing to their similarities in problem structures.

References

1. Blum C, Li X (2008) Swarm intelligence in optimization. Swarm Intelligence: Introduction and Applications. Springer, Berlin Heidelberg, pp 43–85
2. Slowik A, Kwasnicka H (2018) Nature inspired methods and their industry applications-swarm intelligence algorithms. IEEE Trans Industr Inf 14(3):1004–1015
3. Dorigo M, Maniezzo V, Colorni A et al (1996) Ant system: optimization by a colony of cooperating agents. IEEE Trans Syst Man Cybern B Cybern 26(1):29–41
4. Maniezzo V, Colorni A (1999) The ant system applied to the quadratic assignment problem. IEEE Trans Knowl Data Eng 11(5):769–778
5. Bell JE, McMullen PR (2004) Ant colony optimization techniques for the vehicle routing problem. Adv Eng Inform 18(1):41–48
6. Doerner K, Gutjahr WJ, Hartl RF, Strauss C, Stummer C (2004) Pareto ant colony optimization: A metaheuristic approach to multiobjective portfolio selection. Ann Oper Res 131(1–4):79–99
7. Socha K, Dorigo M (2008) Ant colony optimization for continuous domains. Eur J Oper Res 185(3):1155–1173
8. Engin O, Guc¸l¨u A (2018) A new hybrid ant colony optimization algorithm for solving the no-wait flow shop scheduling problems. Appl Soft Comput 72:166–176
9. Dorigo M, Stutzle T (2019) Ant colony optimization: overview and recent advances. in Handbook of Metaheuristics. Springer, pp 311–351
10. Stutzle T, Dorigo M (2002) A short convergence proof for a class of ant colony optimization algorithms. IEEE Trans Evol Comput 6(4):358–365
11. Dorigo M, Gambardella LM (1997) Ant colony system: a cooperative learning approach to the traveling salesman problem. IEEE Trans Evol Comput 1(1):53–66
12. Stutzle T, Hoos HH (2000) Max–min ant system. Fut Gen Comput Syst 16(8):889–914
13. Stutzle T, Lopez-Ib M, Pellegrini P, Maur M, De Oca MM, Birattari M, Dorigo M (2011) Parameter adaptation in ant colony optimization. in Autonomous Search. Springer, pp 191–215
14. Pellegrini P, Sttzle T, Birattari M (2011) A critical analysis of parameter adaptation in ant colony optimization. Swarm Intell 6(1):23–48
15. LeCun Y, Bengio Y, Hinton G (2015) Deep learning XE "Deep learning" . Nature 521(7553):436–444
16. Paszke A, Gross S, Massa F et al (2019) Pytorch: An imperative style, high-performance deep learning library. In: Advances in neural information processing systems 32, H., curran associates, Inc., pp 8026–8037. https://papers.nips.cc/paper/9015-pytorch-an-imperative-stylehigh-performance-deep-learning-library.pdf. Accessed 10 Oct 2020
17. Jia Y, Shelhamer E, Donahue J, Karayev S, Long J, Girshick R, Guadarrama S, Darrell T (2014) Caffe. in proceedings of the ACM International conference on multimedia - MM '14. ACM Press, 2014
18. Ketkar N (2017) Introduction to keras. in Deep learning with python. Apress, pp 97–111
19. Kool W, van Hoof H, Welling M (2018) Attention, learn to solve routing problems. arXiv preprint arXiv:1803.08475
20. Khalil E, Dai H, Zhang Y, Dilkina B, Song L (2017) Learning combinatorial optimization algorithms over graphs. In: Advances in neural information processing systems 30, Curran associates, Inc., pp 6348–6358. https://papers.nips.cc/paper/7214-learningcombinatorial-optimization-algorithms-over-graphs.pdf. Accessed 10 Oct 2020
21. Bengio Y, Lodi A, Prouvost A (2018) Machine learning for combinatorial optimization: a methodological tour d'horizon. arXiv preprint arXiv:1811.06128
22. Klug N, Chauhan A, Ragala R (2019) k-RNN: Extending NNheuristics for the TSP XE "TSP" . Mob Netw Appl 24(4):1210–1213
23. Socha K, Blum C (2007) An ant colony optimization algorithm for continuous optimization: Application to feed-forward neural network training. Neural Comput Appl 16(3):235–247
24. Aljarah I, Faris H, Mirjalili S (2016) Optimizing connection weights in neural networks using the whale optimization algorithm. Soft Comput 22(1):1–15

25. Tian Z, Fong S (2016) Survey of meta-heuristic algorithms for deep learning training. In: Optimization algorithms - Methods and applications. InTech
26. Fong S, Deb S, Yang X (2017) How meta-heuristic algorithms contribute to deep learning in the hype of big data analytics. In: Advances in Intelligent systems and computing. Springer Singapore, pp 3–25
27. Ruder S (2016) An overview of gradient descent optimization algorithms. arXiv preprint arXiv:1609.04747
28. Duchi J, Hazan E, Singer Y (2011) Adaptive subgradient methods for online learning and stochastic optimization. J Mach Lear Res 12(61):2121–2159. https://jmlr.org/papers/v12/duchi11a.html. Accessed 10 Oct 2020
29. Tieleman T, Hinton G (2012) Lecture 6.5-RMSPROP: Divide the gradient by a running average of its recent magnitude. COURSERA: Neural Netw Mach Learn 4(2):26–31
30. Zeiler MD (2012) ADADELTA: an adaptive learning rate method. arXiv preprint arXiv:1212.5701
31. Bottou L (2010) Large-scale machine learning with stochastic gradient descent. In: Proceedings of COMPSTAT'2010, Physica-Verlag HD, pp 177–186
32. Reinelt G (1991) TSPLIB—a traveling salesman problem library. ORSA J Comput 3(4):376–384
33. Cecilia JM, Garc´ıa J.M., Nisbet A., Amos M., Ujaldon M. (2013) Enhancing data parallelism for ant colony optimization on GPUs. J Parall Distribut Comput 73(1):42–51
34. Yang J, Zhuang Y (2010) An improved ant colony optimization algorithm for solving a complex combinatorial optimization problem. Appl Soft Computing 10(2):653–660
35. Elloumi W, Abed HE, Abraham A, Alimi AM (2014) A comparative study of the improvement of performance using a PSO XE "PSO" modified by ACO XE "ACO" applied to TSP XE "TSP" . Appl Soft Comput 25:234–241
36. Olivas F, Valdez F, Castillo O, I Gonzalez C, Martinez G, Melin P (2017) Ant colony optimization with dynamic parameter adaptation based on interval type-2 fuzzy logic systems. Appl Soft Comput 53:74–87
37. Chen J, You X-M, Liu S, Li J (2019) Entropy-based dynamic heterogeneous ant colony optimization. IEEE Access 7:56317–56328
38. Zhou Y, He F, Qiu Y (2017) Dynamic strategy based parallel ant colony optimization on GPUs for TSPs. Sci China Inform Sci 60(6)
39. Skinderowicz R (2016) The GPU-based parallel ant colony system. J Parall Distribut Comput 98:48–60
40. Zhou Y, He F, Hou N, Qiu Y (2018) Parallel ant colony optimization on multi-core SIMD CPUs. Fut Gen Comput Syst 79:473–487
41. Cvetkovic D, Cangalovi M, Kovacevi V (1999) Semidefinite programming methods for the symmetric traveling salesman problem. Integer Programming and Combinatorial Optimization XE "Optimization" . Springer, Berlin Heidelberg, pp 126–136
42. Focacci F, Lodi A, Milano M (2002) Embedding relaxations in global constraints for solving tsp and tsptw. Annal Math Artif Intell 34(4):291–311
43. Xie X-F, Liu J (2009) Multiagent optimization system for solving the traveling salesman problem (TSP). IEEE Trans Syst, Man, Cybern, Part B (Cybern) 39(2):489–502
44. Shi J, Sun J, Zhang Q (2019) Homotopic convex transformation: A new method to smooth the landscape of the traveling salesman problem. arXiv preprint arXiv:1906.03223

Index

W. Li et al. (eds.), *Data Driven Smart Manufacturing Technologies and Applications*, Springer Series in Advanced Manufacturing,
https://doi.org/10.1007/978-3-030-66849-5